PATENT CHALLENGES FOR STANDARD-SETTING in the Global Economy

LESSONS FROM INFORMATION AND COMMUNICATIONS TECHNOLOGY

Committee on Intellectual Property
Management in Standard-Setting Processes

Board on Science, Technology, and Economic Policy

Policy and Global Affairs

Keith Maskus, Editor
Stephen A. Merrill, Editor

NATIONAL RESEARCH COUNCIL
OF THE NATIONAL ACADEMIES

THE NATIONAL ACADEMIES PRESS
Washington, D.C.
www.nap.edu

THE NATIONAL ACADEMIES PRESS   500 Fifth Street, NW   Washington, DC 20001

NOTICE: The project that is the subject of this report was approved by the Governing Board of the National Research Council, whose members are drawn from the councils of the National Academy of Sciences, the National Academy of Engineering, and the Institute of Medicine. The members of the committee responsible for the report were chosen for their special competences and with regard for appropriate balance.

This study was supported by Contract/Grant No. 2010-140-113 between the National Academy of Sciences and the United States Patent and Trademark Office. Any opinions, findings, conclusions, or recommendations expressed in this publication are those of the author(s) and do not necessarily reflect the views of the organizations or agencies that provided support for the project.

International Standard Book Number-13: 978-0-309-29312-9
International Standard Book Number-10: 0-309-29312-X

Additional copies of this report are available for sale from the National Academies Press, 500 Fifth Street, NW, Keck 360, Washington, DC 20001; (800) 624-6242 or (202) 334-3313; Internet, http://www.nap.edu/.

Cover: The cover image of the Rubik's Cube® is used by permission of Seven Towns Ltd. www.rubiks.com. The organizational logos are used by permission of the American National Standards Institute, European Patent Office, European Telecommunication Standards Institute, Institute of Electrical and Electronics Engineers, Organization for the Advancement of Structured Information Standards, United States Patent and Trademark Office and the World Wide Web Consortium which were among the institutions participating in the information-gathering phase of the study. None of these organizations has reviewed or endorsed this report.

Copyright 2013 by the National Academy of Sciences. All rights reserved.

Printed in the United States of America

THE NATIONAL ACADEMIES
Advisers to the Nation on Science, Engineering, and Medicine

The **National Academy of Sciences** is a private, nonprofit, self-perpetuating society of distinguished scholars engaged in scientific and engineering research, dedicated to the furtherance of science and technology and to their use for the general welfare. Upon the authority of the charter granted to it by the Congress in 1863, the Academy has a mandate that requires it to advise the federal government on scientific and technical matters. Dr. Ralph J. Cicerone is president of the National Academy of Sciences.

The **National Academy of Engineering** was established in 1964, under the charter of the National Academy of Sciences, as a parallel organization of outstanding engineers. It is autonomous in its administration and in the selection of its members, sharing with the National Academy of Sciences the responsibility for advising the federal government. The National Academy of Engineering also sponsors engineering programs aimed at meeting national needs, encourages education and research, and recognizes the superior achievements of engineers. Dr. C. D. Mote, Jr., is president of the National Academy of Engineering.

The **Institute of Medicine** was established in 1970 by the National Academy of Sciences to secure the services of eminent members of appropriate professions in the examination of policy matters pertaining to the health of the public. The Institute acts under the responsibility given to the National Academy of Sciences by its congressional charter to be an adviser to the federal government and, upon its own initiative, to identify issues of medical care, research, and education. Dr. Harvey V. Fineberg is president of the Institute of Medicine.

The **National Research Council** was organized by the National Academy of Sciences in 1916 to associate the broad community of science and technology with the Academy's purposes of furthering knowledge and advising the federal government. Functioning in accordance with general policies determined by the Academy, the Council has become the principal operating agency of both the National Academy of Sciences and the National Academy of Engineering in providing services to the government, the public, and the scientific and engineering communities. The Council is administered jointly by both Academies and the Institute of Medicine. Dr. Ralph J. Cicerone and Dr. C. D. Mote, Jr., are chair and vice chair, respectively, of the National Research Council.

www.national-academies.org

COMMITTEE ON INTELLECTUAL PROPERTY MANAGEMENT IN STANDARD-SETTING PROCESSES

Keith Maskus, *Chair,* Professor of Economics, University of Colorado at Boulder
Rudi Bekkers, Assistant Professor of Economics of Innovation and Technical Change, Eindhoven University of Technology, Netherlands
Marc Sandy Block, IP Law Counsel, IBM Corporation
Jorge Contreras, Associate Professor, Washington College of Law, American University
Richard J. Gilbert, Emeritus Professor of Economics, Professor of the Graduate School, University of California, Berkeley
David J. Goodman, Presidential Fellow and Professor Emeritus, Electrical and Computer Engineering, Polytechnic Institute of NYU
Amy Marasco, General Manager for Standards, Strategy and Policy, Microsoft Corporation
Timothy Simcoe, Assistant Professor of Strategy and Innovation, School of Management, Boston University
Oliver Smoot, Past President, International Standards Organization
Richard Suttmeier, Emeritus Professor of Political Science, University of Oregon
Andrew Updegrove, Founding Partner, Gesmer Updegrove, LLP

Project Staff

Stephen A. Merrill, Study Director
Aqila Coulthurst, Program Coordinator
Cynthia Getner, Financial Officer

BOARD ON SCIENCE, TECHNOLOGY, AND ECONOMIC POLICY (STEP)

For the National Research Council (NRC), this project was overseen by the Board on Science, Technology and Economic Policy (STEP), a standing board of the NRC established by the National Academies of Sciences and Engineering and the Institute of Medicine in 1991. The mandate of the STEP Board is to advise federal, state, and local governments and inform the public about economic and related public policies to promote the creation, diffusion, and application of new scientific and technical knowledge to enhance the productivity and competitiveness of the U.S. economy and foster economic prosperity for all Americans. The STEP board and its committees marshal research and the expertise of scholars, industrial managers, investors, and former public officials in a wide range of policy areas that affect the speed and direction of scientific and technological change and their contributions to the growth of the U.S. and global economies. Results are communicated through reports, conferences, workshops, briefings and electronic media subject to the procedures of the National Academies to ensure their authoritativeness, independence, and objectivity.

Paul Joskow (*Chair*), President, Alfred P. Sloan Foundation
Ernst Berndt, Louis E. Seley Professor in Applied Economics, Massachusetts Institute of Technology
Jeff Bingaman, Former Senator, New Mexico, U.S. Senate
Ralph J. Cicerone (ex-officio), President, National Academy of Sciences
Ellen Dulberger, Managing Partner, Ellen Dulberger Enterprises, LLC
Harvey V. Fineberg (ex-officio), President, Institute of Medicine
Alan Garber, Provost, Harvard University
Ralph Gomory, Research Professor, Stern School of Business, New York University
John Hennessy, President, Stanford University
William H. Janeway, Partner, Warburg Pincus
Richard Lester, Japan Steel Industry Professor, Department of Nuclear Science and Engineering, Massachusetts Institute of Technology
David Morgenthaler, Founding Partner, Morgenthaler Ventures
Luis M. Proenza, President and Chief Executive Officer, University of Akron
William J. Raduchel, Independent Director and Investor
Kathryn L. Shaw, Ernest C. Arbuckle Professor of Economics, Graduate School of Business, Stanford University
Laura D'Andrea Tyson, S.K. and Angela Chan Professor of Global Management, Haas School of Business, University of California at Berkeley
Hal Varian, Chief Economist, Google, Inc.
Dan Mote (ex-officio), President, National Academy of Engineering
Alan Wm. Wolff, Senior Counsel, McKenna, Long & Aldridge LLP

Staff

Stephen A. Merrill, Executive Director
Charles W. Wessner, Program Director
Sujai Shivakumar, Senior Program Officer
Paul Beaton, Program Officer
McAlister Clabaugh, Program Officer
David Dierksheide, Program Officer
Aqila Coulthurst, Program Coordinator
Cynthia Getner, Financial Associate

Preface

The U.S. Patent and Trademark Office (USPTO) in 2011 asked the National Academies' Board on Science, Technology, and Economic Policy (STEP) to examine and report on the role of patents in standard-setting processes in an international context. For the STEP program, this charge represented the confluence of its long-standing interests in the standards system on the one hand and intellectual property policy on the other hand. The Board's very first consensus study, in response to a congressional mandate, resulted in the report, *Standards, Conformity Assessment, and Trade* (National Research Council, 1995). And in 2001, STEP initiated a series of studies of the patent system whose products included *Patents in the Knowledge-Based Economy* (National Research Council, 2003), *A Patent System for the 21st Century* (National Research Council, 2004), *Reaping the Benefits of Genomic and Proteomic Research: Intellectual Property Rights, Innovation, and Public Health* (National Research Council, 2006), and *Managing University Intellectual Property in the Public Interest* (National Research Council, 2010). STEP Board recommendations strongly influenced the America Invents Act, enacted in 2011 the first major revision of U.S. patent law in more than half a century.

The present project was approved by the Academies' Governing Board Executive Committee with the following charge:

> An ad hoc committee under the auspices of The National Academies' Board on Science, Technology, and Economic Policy (STEP) will examine and assess how leading national, regional, and multinational standards bodies address issues of intellectual property (IP) arising in connection with the development of technical standards. Through commissioned analysis, a public workshop in Washington and a report of the findings of an expert committee, the project will first document the policies and practices of different types of standard-setting organizations in different geographical contexts. The committee will consider policies with respect to such matters as requirements for the disclosure of IP essential or relevant to the development and implementation of standards, the terms of IP licensing to implementers of a standard, and whether conditions attached to

IP incorporated in standards carry over to a new holder in the event of a transfer of IP rights. The study will assess how these policies work in practice and in a legal context and how variations in these policies relate to different types of standards activities, organizations, and fields of technology. Second, the project will evaluate the effectiveness of these policies in reducing conflict between IP holders and other implementers, balancing the interests of firms of different sizes and with different business models, and balancing the interests of producers and consumers.

A committee comprised of academic economists and social scientists, legal scholars, standards professionals, and technologists was appointed by the Academies to address the charge. The committee met four times in the course of preparing this report. At the first meeting, we received written submissions from or heard oral presentations by individuals from government, industry, and the standards community. We commissioned a study of the IPR policies of a carefully selected sample of national and international SSOs, which was carried out by two members of the study committee, Rudi Bekkers, Eindhoven University of Technology and Andrew Updegrove, Gesmer Updegrove, L.L.P.[1] Next the committee planned and held a two-day symposium, *Management of Intellectual Property in Standard-Setting Processes*, in Washington, D.C., on October 3-4, 2012, with invited presentations on a variety of topics addressed in this report (Appendix B; presentations available at http://sites.nationalacademies.org/PGA/step/PGA_072825). The symposium also provided an opportunity for interested members of the public to express their views. The committee is grateful to all of these contributors to its deliberations.

Our study has been carried out in a dynamic environment. Just in the last few months there have been discussions in numerous SSOs about changes to their IPR policies, new pronouncements from government competition authorities on both sides of the Atlantic, hearings in both houses of Congress, court decisions in high-profile legal suits, and a new articulation of China's policy with respect to "national standards." For the most part, we have taken account of the most important developments through preparation of our report for external review in May 2013. The high profile decision of the United States International Trade Commission in *Apple v. Samsung* that was subsequently overturned by the United States Trade Representative occurred as the committee deliberated its responses to reviewer comments, and the committee could not ignore the relevance of the case's outcomes to its recommendation regarding the availability of injunctive relief to holders of standard-essential patents who have undertaken to license them on reasonable and non-discriminatory terms. Apart from this exception, the committee recognizes that both the intellectual property and standards landscapes are changing and will continue to change in ways that the report does not address.

---

[1] See http://www.nap.edu/catalog.php?record_id=18510.

Preface xi

The committee's recommendations represent a consensus of views, but not every member agrees with every formulation. In one instance, majority and minority views are presented. As with any Academy report, the views expressed are personal and do not necessarily represent the views of members' employers. Despite the heterogeneity of SSOs, the committee's recommendations addressed to standards developers are stated in general terms. The committee recognizes that each organization should and will consider the appropriateness of our advice for its own circumstances and seek its own counsel.

This report has been reviewed in draft form by individuals chosen for their diverse perspectives and technical expertise, in accordance with procedures approved by the National Academies' Report Review Committee. The purpose of this independent review is to provide candid and critical comments that will assist the institution in making its published report as sound as possible and to ensure that the report meets institutional standards for objectivity, evidence, and responsiveness to the study charge. The review comments and draft manuscript remain confidential to protect the integrity of the process.

We wish to thank the following individuals for their review of this report: Alden Abbott, Research In Motion; Andrew Brown, Delphi Corporation; Gary Calabrese, Corning Global Research; Dieter Ernst, East West Center; Patricia Griffin, American National Standards Institute; Irwin Jacobs, Qualcomm; Konstantinos Karachalios, Institute of Electrical and Electronics Engineers Standards Association; Earl Nied, Intel; Joshua Sarnoff, DePaul University; Carl Shapiro, University of California, Berkeley; Andrew Torrance, University of Kansas; and Dirk Weiler, European Telecommunications Standards Institute.

Although the reviewers listed above have provided many constructive comments and suggestions, they were not asked to endorse the conclusions or recommendations, nor did they see the final draft of the report before its release. The review of this report was overseen by Samuel H. Fuller, Analog Devices, Inc. Appointed by the National Academies, he was responsible for making certain that an independent examination of this report was carried out in accordance with institutional procedures and that all review comments were carefully considered. Responsibility for the final content of this report rests entirely with the authoring committee and the institution.

> Keith E. Maskus, *Chair*
> Committee on Intellectual Property
> Management in Standard-Setting Processes
> Stephen A. Merrill, *Study Director*

Contents

SUMMARY 1

1. INTRODUCTION 15
   1.1 Role of Standards and Patented Technology in Standards, 15
   1.2 Standards and Patents in ICT and Emerging Technologies, 18
   1.3 Background of the Study, 19
   1.4 Statement of Task and Organization of the Report, 20
   1.5 Economic Context, 23
   1.6 Standardization in the ICT Setting, 25
   1.7 Stakeholders in Standard-Setting, 28
   1.8 International and Multilateral Issues, 28

2. A COMPARISON OF SSO POLICIES AND PRACTICES 31
   2.1 SSOs Surveyed for the Study, 31
   2.2 A Note on Terminology, 34
   2.3 A Caveat on Coverage, 35
   2.4 SSO Approaches to Basic IPR Issues, 36
   2.5 Transfers of Licensing Commitments, 47
   2.6 Summary Observations, 48

3. KEY ISSUES FOR SSOS IN SEP LICENSING 51
   3.1 Introduction, 51
   3.2 Objectives of FRAND Licensing Obligations, 52
   3.3 Interpretation of FRAND Obligations to Address Competition and
       Efficiency Concerns, 61
   3.4 Recommendations to SSOs, 69

4. SEP DISCLOSURE AND INFORMATION TRANSPARENCY 71
   4.1 Disclosure as an Element of SSO IPR Policies, 71
   4.2 The Possible Roles of Information Disclosure, 72
   4.3 Levels of Disclosure, 74
   4.4 The Timing of Disclosures in Relation to Licensing
       Commitment Procedures, 79
   4.5 Recommendations to SSOs, 80

5. **TRANSFERS OF PATENTS WITH LICENSING COMMITMENTS......81**
   5.1 Introduction, 81
   5.2 Cases Regarding Continuing License Commitments, 83
   5.3 SSO Approaches to Sustaining Licensing Commitments, 88
   5.4 Recommendations for SSOs and Policymakers, 93

6. **INJUNCTIVE RELIEF FOR SEPS SUBJECT TO FRAND......................95**
   6.1 Introduction, 95
   6.2 Views of Competition Regulators, 95
   6.3 U.S. and European Case Law, 100
   6.4 Industry Views, 109
   6.5 Recommendations to SSOs, Courts, and Government Agencies, 111

7. **PATENT OFFICE-SSO COOPERATION 113**
   7.1 Origins and Scope of Information Sharing, 113
   7.2 Benefits and Costs of Information Sharing, 115
   7.3 Legal Status of Standards Information, 116
   7.4 Relevance of the European Experience to the USPTO, 117
   7.5 Recommendations to the USPTO and SSOs, 119

8. **IPR STANDARDS AND EMERGING ECONOMIES 121**
   8.1 Introduction, 121
   8.2 China, 122
   8.3 India, 133
   8.4 Brazil, 137
   8.5 Conclusions and Recommendations, 138

REFERENCES 141

APPENDIXES

A **ACRONYMS..149**

B **SYMPOSIUM AGENDA 153**

C **BIOGRAPHIES OF COMMITTEE MEMBERS AND STAFF................157**

Summary

Background

Standards are technical specifications describing means of achieving certain beneficial features of products and services. To become "standards," such specifications undergo some process of examination and approval, whether through regulatory systems, private industry bodies, or simple market acceptance by consumers, that recognizes they are sufficiently effective to merit wide adoption.

Standards are ubiquitous in today's markets and serve multiple purposes—to assure minimum levels of safety, health, and environmental protection, to provide information to consumers, and to reduce transaction costs between producers and users in the selection of inputs and products. One of the most important functions of contemporary standards, and the focus of this report, is to enable components and products designed and produced by different firms to operate and communicate with one another. Such interoperability standards are increasingly important for domestic and international commerce by helping to achieve economies of scale and scope within and across borders.

The technologies that enter into standards are often protected by patents or are the subject of patent applications at the time standards are developed. Incorporating patented or patent-pending technologies in standards is virtually inevitable and generally beneficial, but there is a tension between owners and users of a patented technology. Inventors generally seek economic returns on their R&D investments while users of technologies want access to them on affordable terms. This tension is even more pronounced in the realm of standards, which by their nature are intended to have widespread acceptance and use.

To manage this tension, the wide variety of entities, domestic and international, that are dedicated to developing standards (termed "standard-setting organizations" or SSOs in this report) have generally adopted policies regarding the disclosure and terms of licensing of patents essential to the standards they create (so-called standard-essential patents or SEPs). In general, SSOs encourage or require member firms to disclose SEPs and license them to standards implementers under terms commonly referred to as fair, reasonable, and non-discriminatory (FRAND). These policies vary in content and specificity, are in many cases in flux, and often lack guidance for increasingly common occurrences—litigation over SEPs and changes in SEP ownership. In particular, SSO policies often do not address whether a SEPs holder that has made a FRAND commitment should be able to seek injunctive relief or an order excluding the allegedly infringing product from the United States and whether FRAND licensing commitments by patent holders in an SSO transfer with changes in patent ownership.

At the same time that the voluntary standards development system common in most respects to the United States, Europe, and Japan is evolving, it is also adjusting to the rise of large developing economies that are major markets for new technologies and show promise of becoming important sources of them. There is uncertainty about how standards policies will evolve in China, India, and Brazil in particular and how they will treat intellectual property incorporated in standards. In a world of rapid technological change and diffusion, proliferating patents, and frequent litigation over patents, the relationship of patents to standards obviously has enormous implications for firms, national economies, and global trade.

Study Origin, Methods, and Focus

In 2011, the U.S. Patent and Trademark Office (USPTO) asked the National Academies to examine and report on the role of patents in standard-setting processes in an international context. The Academies appointed a committee composed of academic economists and social scientists, legal scholars, standards professionals, and technologists and charged them with documenting and evaluating the policies and practices of different types of SSOs in different geographical contexts, focusing on such matters as patent disclosures, terms of licensing, and provisions for the transfer of obligations when patents are traded, sold, or disposed in bankruptcy proceedings.

The committee held four meetings, including a workshop with presentations selected by the committee as well as public commentary and commissioned original research and analysis, including a study of a dozen SSOs operating in the information and communications technology (ICT) sector. The committee, in consultation with the sponsor, chose ICT as the project's focus because of its technological dynamism and heavy reliance on standardization, and because of the escalation of patenting and salience of issues involving patents and standards in those industries.

SSO Approaches to IPR Issues

The committee's selection of SSOs to examine represents a diversity of organization types (both formal standards organizations and consortia) and geographical foci (U.S., European, and global) and encompasses standards activity across the range of ICT technologies—consumer electronics, microelectronic products and their associated software and components, and communications networks including the Internet. These organizations and their salient characteristics are listed in Table S-1.

TABLE S-1 Organizations and their Salient Characteristics

TITLE	TYPE	GEOGRAPHICAL FOCUS	TECHNOLOGY FOCUS	NOTABLE IPR POLICIES
International Organization for Standardization (ISO)	formal	global	broad	share common policy but permit adjustments
International Telecommunications Union (ITU)	formal	global/UN Affiliated	communications	
International Electrotechnical Commission (IEC)	formal	global	electrical, electronics-related technologies	permits but does not require *ex ante* disclosure of the terms
Institute of Electrical and Electronics Engineers Standards Association (IEEE-SA)	formal professional association	global	broad electronics	reviews ipr policies as part of accreditation process
American National Standards Institute (ANSI)	SSO and standards-accreditation organization (not a standards developer)	U.S.	broad	
Internet Engineering Task Force (IETF)	consortium of individuals	global	internet	preference for non-patented technology
Organization for the Advancement of Structured Information Standards (OASIS)	consortium	global	e-business and web service	multi-modal ip policy
VMEBus International Trade Association (VITA)	consortium	global	avionics, military and industrial applications of electronics	*ex ante* disclosure of licensing terms; binding arbitration of disputes
World Wide Web Consortium (W3C)	consortium	global	internet & web	royalty free license
High Definition Multimedia Interface Forum (HDMI)	consortium	global	digital audio/visual transmission	non-assertion
Nearfield Communications Forum (NCF)	consortium	global	data exchange among consumer devices	
European Telecommunications Standards Institute (ETSI)	formal	European-based but international	ICT broadly	

All of these standard-setting organizations have a diverse set of stakeholders and constituents. Some participants are technology owners whose business models depend on sales of products or services, less so or not at all on royalties for SEPs, although they may want their patents to have sufficient value to offset the rights held by others in the same technology area. Other participants are technology sellers whose models are based on royalties from implementers for SEPs and even non-SEPs. Others are technology users seeking low or no royalties for the SEPs they license from others. And still others are both technology users and sellers, who may assume different postures in different standards-setting processes. SSO IPR policies are shaped both by the interests of existing members and by the need to attract new participants who may be technology sellers, users, or both. This divergence of interests and the difficulty of reconciling them may account for the fact that very few SSO articulate a clear set of objectives for its IPR policies, making it difficult to evaluate their effectiveness.

For most SSOs, however, the minimum goal of their IPR policies is to ensure that all essential patent claims are reasonably known to the participants and are available for licensing under a FRAND or a similar framework minimizing the potential for *ex post* hold-up and royalty stacking. Beyond that there is wide variation and often considerable ambiguity in the rules regarding **Disclosure:**

- Whose patents must be disclosed; what qualifies as an "essential" patent or patent claim; when disclosures must be made in the standards development process; whether blanket (non-patent specific) disclosures suffice; to whom the disclosed information is provided; and whether there is a requirement to update disclosures, for example, as a standard evolves and as patents are issued or denied.

SSO policies regarding **Licensing** are, if anything, even more varied and in some instances ambiguous:

- What specific terms or limitations are imposed by a commitment to FRAND licensing; what is meant by the individual terms "fair," "reasonable," and "non-discriminatory"; whether a maximum royalty must be posted before the standard is adopted ("*ex ante*"); how FRAND applies to portfolio licenses and cross-licenses; how non-royalty licensing terms (e.g., grant-backs, geographical or field of use limitations, etc.) are treated; and whether royalty-free licensing is encouraged or required.

Inasmuch as patents and patent portfolios are now more frequently traded, sold, or acquired through bankruptcy, it is becoming increasingly important and complex for SSOs to address the **Transfer** issues surrounding FRAND-encumbered SEPs. Concerned that standard implementers could be at risk of hold-up by a new SEP owner, competition authorities have generally taken the position that SSOs should create contractual commitments that, subject to local law, bind successors to original FRAND obligations. Some SSOs have taken or are in the process of taking such steps but the legal issues are complex.

On the other hand, few SSOs have addressed the controversial question of whether and how a FRAND commitment should affect a SEP owner's ability to seek or threaten to seek **Injunctive Relief** (or, in the case of an imported product, an exclusion order by the U.S. International Trade Commission) as a remedy for patent infringement. Again, competition authorities are generally agreed that in the case of a FRAND-encumbered SEP an injunction should be an infringement remedy of last resort. But industry views are divided, as might be expected; and as a result, none of the SSO policies the committee examined imposes any restrictions on what legal remedies a member or third-party beneficiary of a licensing commitment may pursue in court or in the International Trade Commission.

Recommendations for SSO and Government Policies

Having studied the experience of the dozen SSOs examined in the commissioned paper, the positions of government regulators, the evolving case law in areas of legal uncertainty and contention, and economic theory, the committee recommends that SSOs consider a number of policies with regard to intellectual property, while recognizing the diversity of stakeholder interests and their variation from organization to organization. In some cases, the committee also recommends actions by government authorities supportive of these principles.

Interpretation of FRAND

The committee believes that a FRAND licensing commitment represents more than the patent owner offering a license on terms of its own choice. A FRAND commitment is also mutual in the sense that both the SEP holder and any prospective licensee are expected to negotiate in good faith towards a license on reasonable terms and conditions that reflect the economic value of the patented technology.

Recommendation 3:1[1]

The committee urges SSOs to become more explicit in their IPR policies regarding their understanding of and expectations about FRAND licensing commitments. SSOs should clarify the various effects of a FRAND commitment by formulating certain statements of principle. These principles could include, among other conditions for compliance with FRAND, guidance regarding royal-

---

[1] The first numeral of each recommendation indicates the chapter of the report in which it is discussed.

ty demands that could be a disproportionate share of product value when many patents are necessary to comply with a standard and the relevant product includes multiple technologies.

Recommendation 3:2

The committee recommends that SSOs include statements in their policies that implementers and the consumers of their products and services are the intended third party beneficiaries of licensing commitments made by SSO participants. Although the enforceability in all courts of such a term may not be guaranteed (the law in this regard is still evolving), inclusion of such statements would inform courts of the intent of SEP owners participating in SSO working groups. It would also provide greater confidence to potential implementers, and promote greater certainty in the event of a dispute.

Several recommendations are aimed at improving clarity within SSOs regarding the bundling of licensing commitments.

Recommendation 3:3

SSOs should clarify in their policies that prospective licensees may request a license to some or all FRAND-encumbered SEPs owned or controlled by a patent holder. Licensors may not tie the FRAND commitment and the availability of the requested SEPs to a demand that a licensee accept a package or portfolio license that includes non-SEPs or SEPs for unrelated standards. Nor may the licensors tie the FRAND commitment and SEPs availability to a requirement that the licensee agree to license back unrelated SEPs or non-SEPs.

Recommendation 3:4

SSOs should clarify in their policies that a holder of FRAND-encumbered SEPs may require a licensee to grant a license in return under FRAND terms to the SEPs it owns or controls (and those of its affiliates as specified in the SSO's policy) covering the same standard or, as specified by the SSO, related standards.

Recommendation 3:5

It should be understood that SSOs' IPR policies do not affect the freedom of parties to voluntarily enter portfolio or cross licenses beyond the scope of the standard. This includes situations where prospective licensors offer to license SEPs in a package, such as a fixed pool.

Patent Disclosures

The committee recognizes that many aspects of transparency are subject to tradeoffs, not only for SSOs but also for member companies. On the one hand, more transparency can reduce uncertainty and legal exposure and can be beneficial in cases of conflict. On the other hand, achieving transparency through disclosures can involve significant effort and compliance costs. The committee nevertheless recommends that SSOs consider the following steps to increase transparency of SEP ownership and licensing.

Recommendation 4:1

SSOs that do not have a policy requiring FRAND licensing commitments from all participants should have a disclosure element as part of their IPR policy.

Recommendation 4:2

SSOs with disclosure policies should articulate their objectives and consider whether they sufficiently serve these objectives. In particular, such SSOs may consider separating patent disclosure from licensing commitments and better define their preferred timing and specificity of disclosures.

Recommendation 4:3

SSOs should make disclosed information available to the public.

Recommendation 4:4

SSOs should consider measures to increase the quality and accuracy of disclosure data. Such measures might include updating requirements or greater coordination with patent offices.

Transfer of Patents with Licensing Commitments and Transparency of Patent Ownership

Transfers of standard-essential patents are an increasingly important feature of the high-technology marketplace, as a result both of firms seeking to realize their economic value through sales of assets being disposed in bankruptcy proceedings. Such transfers raise complex issues regarding the obligations and rights of transferors, transferees, and existing and potential licensees along what may be an extended chain of transactions. Statutes and judicial rulings so far provide at best partial guidance, and there are significant differences in law across countries. Major competition authorities, on the other hand, see clear value in binding transferees to original commitments.

The committee agrees with these authorities that a FRAND commitment should travel with the patent when it is transferred, although there are different means and modalities by which that could occur. Satisfactory resolution of the complex issues of patent transfers may require government action and possibly legislation. First, U.S. law does not require recordation of ownership changes or the identity of real parties of interest with the U.S. Patent and Trademark Office. In the committee's judgment, lack of transparency is no longer acceptable in an era of vibrant markets in intellectual property and frequent bankruptcy proceedings in which patents are a principal asset. Registration of SEP ownership changes with SSOs would represent only a partial solution and could become burdensome to some of these organizations.

The committee makes the following suggestions for SSO policies and public policies to advance that proposition and minimize uncertainty and additional transaction costs for licensees.

Recommendation 5:1

Where they have not already done so, SSOs should develop meaningful policies by which successors in interest are bound to whatever licensing commitment (e.g., FRAND) the SEP owner made to the SSO in question under that organization's IPR policy. This requirement should apply whether SEPs are individually disclosed or are covered by a blanket disclosure. These obligations should cascade through succeeding transfers.

Recommendation 5:2

Legislation, case law, or other legal mechanisms should tie licensing commitments to FRAND-encumbered patents needed to implement SSO standards. This should be done in ways that ensure the commitment automatically runs with the patents.

Recommendation 5:3

It may be difficult to identify patent transfers, because under current U.S. law they need not be recorded. Accordingly, public recordation with the patent office of transfers of all patents should be required by legislation or regulation. The committee believes that this approach of recording all patent transfers is a practical and effective way of enabling transparency for transfers of SEPs, which may not always be identified as such. The record should identify the real party in interest.

Recommendation 5:4

Bankruptcy concerns are especially complex and raise uncertainty about consistency of licensing commitments. SSOs should develop guidelines to ensure that the licensing assurances made to them remain with the patent in bankruptcy proceedings and support legislation, if necessary, to the same end.

Recommendation 5:5

Competition authorities and international policy negotiators should, through legislation or regulation, find means to reduce inconsistencies across national legal jurisdictions in patent-transfer issues, including in bankruptcy processes.

Injunctive Relief for SEPs Subject to FRAND

The committee believes that a FRAND commitment limits a licensor's ability to seek injunctive relief, including exclusion orders, and recommends the following steps to help avoid or resolve disputes, prevent anti-competitive conduct, and ensure reasonable compensation to SEP holders whose patents are infringed.

Recommendation 6:1

SSOs active in industries where patent holdup is a concern should clarify their policies regarding the availability of injunctions for FRAND-encumbered SEPs to reflect the following principles:

- Injunctive relief conflicts with a commitment to license SEPs on FRAND terms and injuctions should be rare in these cases;
- Injunctive relief may be appropriate when a prospective licensee refuses to participate in or comply with the outcome of an independent adjudication of FRAND licensing terms and conditions; and
- Injunctive relief may be appropriate when a SEP holder has no other recourse to obtain compensation.

The committee could not reach unanimous agreement on appropriate venues for adjudicating FRAND disputes. However, a majority of the committee members endorse the following:

Majority Recommendation 6:2

SSOs should clarify that disputes over proposed FRAND terms and conditions should be adjudicated at a court, agency, arbitration or other tribunal that can assess the economic value of SEPs and award monetary compensation.[2]

---

[2] A minority of committee members endorse this alternative recommendation: Courts, agencies, arbitration bodies or other tribunals (including the USITC) that consider patent essentiality, FRAND determination, or public interest factors should be presented with

The committee also could not reach unanimous agreement on the scope of any limitations that a FRAND commitment might place on SEP holders' rights to seek injunctive relief. However, a majority of the committee members endorse the following recommendation in that regard:

Majority Recommendation 6:3

SSOs should clarify that, before a SEP holder can seek injunctive relief, disputes over proposed FRAND terms and conditions should be adjudicated at a court, agency, arbitration, or other tribunal that allows either party to raise any related claims and defenses (such as validity, enforceability and non-infringement).[3]

SSO-Patent Office Information Sharing

As the interplay between standards and patents has increased, so has recognition that the functioning and integrity of the two systems are interdependent. Up-to-date information on claims in issued patents and on the status of patent applications can be very useful to standards development working groups. Likewise, the submissions of participants to standards bodies as well as finalized standards documents represent a potentially valuable collection of prior art for consideration by patent examiners. All parties have a stake in the quality of issued patents.

The European Patent Office has agreements with three SSOs to share such information in standardized format. The agreements with the International Telecommunications Union (ITU), European Telecommunications Standards Institute (ETSI), and Institute of Electrical and Electronics Engineers-Standards Association (IEEE-SA) have these common elements: exchange of information and documentation; collaboration on documentation format and dissemination policies to align them with EPO prior art search needs; standards-education of EPO personnel; and self-funding of the expenses involved. Participants in these arrangements agree that their value extends beyond generating patents of higher quality. Improved transparency in the linkages between IPR and standards is seen as a benefit by both SSO members and EPO examiners.

The committee finds that arrangements along the lines of the EPO-SSO memoranda could benefit both the U.S. Patent and Trademark Office examina-

---

the facts and render injunctive relief decisions based on existing law, such as the *eBay* decision and/or ITC Section 337.

[3] A minority of committee members endorse this alternative recommendation: SSOs should clarify that a SEP owner that has made an offer and offered to negotiate, with a prospective licensee, a license that will embody FRAND terms should be allowed to include injunctive relief in its pleadings when a FRAND dispute is brought to a court, agency, arbitration, or other tribunal that can consider equities, party conduct, reciprocity, and FRAND factors (including FRAND rates and terms).

tion process and SSO functioning at relatively modest cost. That would likely entail agreements separate from the EPO's with ITU, ETSI, and IEEE-SA. The committee did not fully explore two legal issues raised in its discussions: 1) how in the wake of the America Invents Act the legal definition of prior art varies across jurisdictions and between the EPO and the USPTO in particular; and 2) whether the form of the USPTO's cooperation with individual SSOs could represent a conflict of interest for the agency. The committee recommends that the USPTO take the following steps:

Recommendation 7:1

In the wake of the passage and implementation of the America Invents Act, the USPTO should

- Clarify how the legal definition of prior art varies across jurisdictions, particularly as between the EPO and USPTO. Specifically, when is art "publicly available" in a standards context?
- Explore with leading SSOs, including possibly ETSI, IEEE-SA and ITU, information-sharing arrangements similar to those concluded by the EPO;
- Work with other patent offices to establish uniform fields and templates for standards-based prior art documents, such as early drafts of specifications, published minutes, and the like and deliberate with other offices on the definition of sharable information in this context;
- Improve standards technology education for U.S. patent examiners. For example, when standards developers convene in Washington, D.C., they could be asked to instruct and update USPTO examiners about standards processes and recent developments; and
- Develop joint education programs with SSOs on the pros and cons of standards-based prior art, especially early drafts, and the benefits from including it in patent office search databases.

Standards Processes and Policies in China, India, and Brazil

Although a larger number of countries are significant players in global ICT markets—among them Korea, Taiwan, Hong Kong, and Singapore—the USPTO asked the committee to focus its study on three emerging players for the following reasons. China, India, and Brazil represent three very large, rapidly growing economies that, until recently, have been reliant on imported technologies subject to the standards developed in the institutions described above which are dominated by the United States, Europe, and Japan. But in all three cases national governments are making significant new industrial policy commitments intended to foster national innovation capabilities and push their economies into higher value-added, more knowledge intensive production. Their approaches to standards development and IPR should be seen in the context of these broader industrial policy goals. But they are also conditioned by membership in the World Trade Organization (WTO) and adherence to the Agreement on Technical Barriers to Trade (TBT) and Agreement on Trade-Related Aspects of Intellectual Property Rights (TRIPS), which compel a degree of harmonization with international norms. In all three countries, policies and institutions are in transition or development and could take a more nationalistic turn. Given the size of these economies, how these policies develop will have important implications for the norms and institutions by which international standardization and IPR affairs evolve.

China

Chinese approaches to standardization, including policies for essential patents in standards, have evolved in ways that show sensitivity to international norms and the interests of international stakeholders, including multinational corporations, in some circumstances, although in other cases have excluded foreign companies' participation. At the same time, the government is strongly committed to building a standards regime to serve Chinese national interests, including reducing dependence on imported technology and the associated licensing fees and building domestic innovation capacity. This is evident in efforts over the past decade by the Standards Administration of China (SAC) to formulate national policy guidelines, most recently the "Regulatory Measures on National Standards Involving Patents (Interim)." These guidelines endorse disclosure and FRAND licensing norms but leave a number of definitional and procedural ambiguities that suggest a bias against the interests of rights holders. Meanwhile, China is experiencing a huge increase in patent litigation, some of it involving SEPs-related cases.

India

In contrast to China, India has shown less of a strategic orientation to standards development and intellectual property and more inclination to follow the policies for IPRs in standards of the established international standards bodies. However, there are signs of a growing appetite for government-supported indigenous Indian standards development efforts incorporating Indian intellectual property. In particular, the 2012 National Telecom Policy calls for numerical targets for the growth and self-sufficiency of the industry and the creation of "… a roadmap to align technology, demand, standards, and regulations for enhancing competitiveness of domestic manufacturing…" through establishment of "standards to meet national requirements, generate IPRs, and participate in international standardization bodies… making India a leading nation in the area of international telecom standardization."

Brazil

Like India, Brazil has shown less of a strategic orientation towards standards and IPR development and its standards institutions, lacking well-developed policies for dealing with IPR in standards, are only beginning to come to grips with the complex issues involved.

In all three countries reviewed, the development of a modern technical standards regime is still a work in progress. This is true even for China, where its learning curve is notably steep and it has shown a far more robust approach to building a national standardization system than has India or Brazil. While there are limits to how much the U.S. government can contribute to the development of these standards regimes, the fact that they are all in varying stages of formation suggests that there are possibilities for mutually beneficial interactions, especially with regard to education, training, and raising awareness on the importance of developing IPR policies in the early stages of building SSO capabilities.

Recommendation 8:1

The U.S. government should explore ways to promote awareness of the importance of developing IPR policies at an early stage of the development of SSOs in these and other emerging economies, and should, in conjunction with non-governmental standards entities, explore ways to offer training programs for those working to develop their organizations and policies needed for successful national standardization.

Recommendation 8:2

In the meantime, the relevant agencies of the United States government, such as the United States Patent and Trademark Office, the Office of the United States Trade Representative, and the National Institute of Standards and Technology, should closely monitor and report on continuing developments in these countries and other major emerging economies regarding standard-setting and the management of intellectual property.

1. Introduction

1.1 The Role of Standards and Patented Technology in Standards

This report addresses the complex relationships among the development of technical standards, their ownership, in part as intellectual property (primarily in the form of patents), and their diffusion into competition. For reasons described below, the report focuses on the information and communications technology (ICT) industry where issues involving these relationships especially rise to the fore because of the importance of licensing essential patented technologies across multiple uses.

Standards are technical specifications that aid the development of certain beneficial features of products and services. To become "standards," such specifications undergo some process of examination and approval, whether through regulatory systems, private industry bodies, or simple market acceptance by consumers, that recognizes they are sufficiently effective to merit wide adoption. Standards are ubiquitous throughout markets and are adopted for multiple purposes. For example, they exist in agriculture, foodstuffs, and medicines to assure minimum safety and health levels. Emissions standards apply to electricity generation and automobiles to improve air quality, while banks are subject to fiduciary requirements to safeguard financial stability. Standards determine minimum levels of information that must be provided to the public by government agencies and commercial enterprises. Standards also play a useful signaling function, for their adoption signifies compliance with specified performance characteristics.

Standards are developed to resolve various market shortcomings in unregulated markets. These problems may arise from externality costs, such as pollution of air and water, lack of information about the health and safety characteristics of goods and services, and network vulnerabilities in the electrical grid and financial markets. They also reduce transaction costs between technology producers and users in the selection of inputs and products. Without question, the development and adoption of appropriate standards to address such issues help create markets, support the functioning of efficient competition, and raise consumer welfare.

Standards have taken on increasing importance for international commerce. For example, empirical evidence points to the important role standardization plays in supporting international trade by raising consumer confidence in traded goods (Moenius 2004; Clougherty and Grajek 2012). They also support increasing global investment by facilitating information sharing and data transfers among affiliates within multinational firms and among participants in international research networks. In turn, the use of standardized technologies helps achieve economies of scale and scope, both within and across borders. Innovation and growth are increasingly dependent on the development and use of appropriate standards.

Many standards, including those of greatest relevance for this report, enable products designed and produced by different companies to operate and communicate with one another. Such "interoperability" standards, when implemented broadly across markets, give rise to beneficial network effects and efficiencies. Interoperability standards are important in many industries but particularly characterize the information technology, mobile telephone, and consumer electronics sectors. Indeed, standards can effectively create new markets in such sectors. The positive implications for market efficiency and consumer welfare are clear: the world would be more fragmented and unproductive if software and telecommunication technologies could not operate across multiple platforms and devices.

Technology developers often rely on patents to commercialize their inventions and, ultimately, to support investments in research and development. These investments often produce technologies that are incorporated into standards. Indeed, in many fields of ICT, a substantial share of relevant technology is patented or the subject of patent applications at the time a standard is developed. Thus, the incorporation of patented technologies into ICT technical specifications is virtually inevitable and, by facilitating the benefits of standardization, is ultimately in the public interest. Incorporating these patented inventions can result in a standard with better performance, improved cost effectiveness, or a better match with other design requirements. It is increasingly the case in ICT that some design requirements cannot be met at all without including patented technology. Moreover, the potential to receive royalties for access to patented technology creates incentives for participation in standard-setting and attracts parties that contribute valuable knowledge and technical insights.

There is an inherent tension between the interests of inventors, who seek economic returns on their R&D investments, and users of new technology, who want access on affordable terms. This tension is even stronger in the area of standards, which, by their nature, must find widespread acceptance and, if patented, may give rise to two problems for market competition. The first is lock-in, where patented technology that is not readily replaceable must be implemented for products to work. The second is potential hold-up, where patent holders seek royalties substantially in excess of the value a technology had prior to its incorporation into a standard.[1] In this context, policies and guidelines governing how widely and the terms under which such technologies are licensed are critical for supporting markets in downstream products.

---

[1] See Chapter 3 for extensive discussion.

Introduction

Striking a sensible and efficient balance regarding the management and licensing of intellectual property, primarily standard-essential patents (SEPs), in ICT standards is a central problem for standard-setting organizations (SSOs), their members, and government authorities. The importance of balance is especially pronounced where there is a critical need for seamless interoperability among software, components and other technologies embedded in microelectronic devices, such as cellular telephones, and other technically sophisticated products, and where there are a large number of patents on the relevant inputs. Owners of patents on technologies to which access is required for making these products work certainly have incentives to license their use to implementers of various complementary technologies and products. However, the licensing terms and conditions they set may limit the access of some potential licensees who could otherwise bring successful products to market or may impose costs that impede technology utilization.

It should be noted that there are significant amounts of innovation, including of technologies that enter standards, from open-source approaches. Open-source innovation is common in software and certain segments of microelectronics and biotechnological research. Such approaches preclude the assertion of patent rights on new technologies, though there may be other restrictions on licensed use. Open-source is an increasingly relevant source of knowledge for standardization. However, it does not raise the same questions that this report addresses, namely the management of intellectual property rights in the standardization process.

Standard-setting organizations (SSOs) play several critical roles in technology and market development. The ultimate objective is to enable competition among rival but interoperable products and services without permitting some participants to block others by inefficiently asserting patents. Thus, their first important task is to ensure the interoperability of technology products and to facilitate the necessary exchange of data through the development of industry standards. Second, these organizations realized long ago that to foster competition they needed to place controls on how member firms manage their SEPs. Thus, they developed intellectual property rights (IPR) policies intended to ensure reasonable access to patented technology necessary to implement their standards.

How SSOs operate varies widely across technologies and regions, as this report will demonstrate. In general, however, they seek to encourage or require member firms to both disclose and license SEPs, whether to fellow SSO members or non-member companies, under terms generally referred to as reasonable and non-discriminatory (RAND) or fair, reasonable and non-discriminatory (FRAND).[2] The purpose of the FRAND framework is to facilitate the licensing

---

[2] What these terms might actually mean is a subject of considerable discussion and controversy, as will be discussed later in the report. It is generally agreed that the terms may be used interchangeably. FRAND is used in this report because it is generally used

of critical patented technologies to designers and implementers of components and final goods. Where multiple technologies may be needed to ensure interoperability and functionality but are not available, competition, innovation, and the growth of markets may be stunted. For this reason, how firms and SSOs approach licensing of SEPs and other patents recently has attracted the attention of competition authorities in major jurisdictions.

Many complex questions arise in this arena. For example, what are the IP disclosure expectations of SSOs and are they mandatory rules or voluntary guidelines? Do SSOs define various licensing practices and requirements and do such practices vary between member and non-member users and implementers? Is there a common understanding of the effects of a FRAND licensing commitment? Are there limitations regarding the role of SSOs in defining or enforcing licensing obligations? How do such procedures and policies vary across geographic regions, including in major emerging economies? Finally, although the committee does not address this question directly, how does the recent mushrooming growth of patents in key industries affect the ability of SSO members to manage and license their intellectual property to facilitate technology use?

1.2 Standards and Patents in ICT and Emerging Technologies

This report focuses on such questions in the ICT sector because interoperability needs and network economies are critical in this area. Moreover, patenting has become more prevalent in this sector in multiple countries, and those patents protect many of the technologies written into standards. Firms seeking to implement standards or develop improved technologies would necessarily infringe the patents embodied in those standards unless they have legal access through licensing. The associated questions of selection and disclosure of essential patents included in standards and the terms for licensing them are especially salient in ICT.

Despite this focus, the committee invited presentations at its workshop on several emerging technologies, including bioinformatics, synthetic biology, nanotechnology, and sustainable (green) building materials.[3] The presenters described some parallels with ICT in these fields but reported that, to date, the complex patent issues raised in the ICT field have yet to attract a significant attention in these other fields. For example, the field of bioinformatics has evolved largely within academic and governmental research centers. Numerous standards have been developed for data structures and exchange, primarily in small, academically focused groups. Patent issues have not received much, if any, attention and it does not ap-

---

internationally. Of course, FRAND is not the only licensing framework entertained by SSOs, as noted in the next chapter.

[3] See, respectively, invited presentations from Contreras (2012), Torrance and Kahl (2012), Jillavenkatesa,et al (2012), and Contreras and McManis (2012). http://sites.national academies.org/PGA/step/IPManagement/index.htm.

Introduction

pear that patenting has yet occurred with any frequency in this field, although the potential for filings in some subfields such as genetic data structures could increase in the future.

Synthetic biology, which also originated in the academic research environment, has attracted the interest of private sector players, who have been active in patenting synthetic biology inventions. Although the potential need seems substantial, few standards have been developed in the field. The most prominent standardization effort to-date, the "bio-bricks" project that has developed a large and growing catalog of standardized molecular "parts," has sought to discourage patenting of these fundamental molecular elements. Nanotechnology is a more mature field and has numerous private sector players. To the extent that standards are being developed, this activity is occurring at large, established SSOs such as American Society for Testing and Materials (ASTM) International and ISO, which have patent policies in place.

Finally, sustainable building materials represent a large commercial market with numerous sophisticated players. A large number of standards have been developed at a range of SSOs, from small, industry-specific trade groups to large SSOs such as Underwriters Laboratories. In this field, trademarks and certification marks (so-called "ecolabels") have played a far greater role than patents, and present their own challenges to participants, regulators and consumers.

Despite the relatively low salience of patent issues to date, they have the potential to assume greater importance in each of the fields in the coming years. Accordingly, the committee believes that its findings and recommendations with respect to ICT may have some value to participants seeking to anticipate issues that may be problematic in the future.

There are many long-established SSOs, many with IPR policies operating in relatively mature sectors of the economy such as automobiles and aerospace (e.g., Society of Automotive Engineers), electrical machinery (e.g., National Electrical Manufacturers Association), and other manufacturing industry groups (e.g., Society of Manufacturing Engineers). These are not considered in this report and there should not be any inference that the committee's findings and recommendations apply equally to such industries and organizations.

1.3 Background of the Study

In 2011, the United States Patent and Trademark Office (USPTO) commissioned the Board on Science, Technology and Economic Policy (STEP) of the National Academy of Sciences to empanel a committee of experts to study and prepare a report analyzing such questions, with an emphasis on an international comparison. The committee began its work in November 2011, with an initial meeting at which stakeholders and other interested parties were invited to make statements on the issues. Subsequently, the committee asked a number of experts to prepare presentations addressing specific aspects of SSOs, licensing and related issues. The presentations were made at a public symposium in October 2012. In addition, members of the committee engaged in extensive discussions over the period to prepare for the drafting of this report.

A principal motivation of the USPTO request is the increasing importance of standardization and IPR use for global trade and investment. In one dimension, this reflects the growing prevalence of cross-border activities of SSOs, raising the question of how they manage their IPR policies in an international context. In another context, major emerging economies, especially China, now place greater emphasis on their own standardization bodies and associated policies. Thus, the committee was asked to ascertain the status of standardization in high-technology areas in such economies.

Another important context of the study is the growing concern that in some high-technology sectors the system faces increasing difficulties in effectively disseminating the use of patented technologies in key standards. The concern derives in part from the proliferation of high-stakes patent lawsuits involving SEPs in many countries and requests for injunctions to exclude alleged patent infringers from various national markets. Such episodes raise the question of whether the current system strikes an appropriate balance among the various stakeholders in standards and IPR.

There are many indications of the current prominence of these issues. All three branches of the United States government and the European Union have taken an active interest in the relationship between standards and patents. Since our committee began its work there have been hearings in the House of Representatives and the Senate, appellate and district court and International Trade Commission decisions, interventions by the Federal Trade Commission and two Statements of Objections by the Directorate General for Competition in the European Commission, as well as policy statements issued by the Department of Justice, the Federal Trade Commission, and the USPTO, all involving disputes over SEPs and the effect of a FRAND commitment. In this context, one impetus for the study was a desire for an independent, expert view on how public agencies might contribute better to the evolving standards environment to promote competition and growth.

1.4 Statement of Task and Organization of the Report

The Statement of Task agreed to by the USPTO and the National Academies directed the committee to:

> ...examine and assess how leading national, regional, and multinational standards bodies address issues of intellectual property (IP) arising in connection with the development of technical standards. Through commissioned analysis, a public workshop in Washington and a report of the findings of an expert committee, the project will first document the policies and practices of different types of standard-setting organizations in different geographical contexts. The committee will consider policies with respect to such matters as: requirements for disclosure of IP essential or rel-

Introduction

evant to the development and implementation of standards, the terms of IP licensing to implementers of a standard, and whether conditions attached to IP incorporated in standards carry over to a new holder in the event of a transfer of IP rights. The study will assess how these policies work in practice and in a legal context and how variations in these policies relate to different types of standards activities, organizations, and fields of technology. Second, the project will evaluate the effectiveness of these policies in reducing conflict between IP holders and other implementers, balancing the interests of firms of different sizes and with different business models, and balancing the interests of producers and consumers.

With this statement in mind, the committee set the following objectives for its report:

- Identify and survey a representative selection of major SSOs, with operations in different major countries or with members from multiple regions, in order to describe and document their relevant objectives, policies and practices. The results of this survey are summarized in Chapter 2 of the report.
- Develop economic and legal analysis of critical issues surrounding management of intellectual property in SSOs, including the use and meaning of FRAND licensing. This discussion is in Chapter 3.
- Consider how the policies of SSOs relate to obligations or expectations regarding the disclosure of essential IPR and commitments to license them. This is the subject of Chapter 4.
- Review the implications when FRAND-encumbered SEPs are transferred. Issues analyzed include how the FRAND assurance made by a SEP owner is addressed after the transfer, how parties' rights are affected by the exchange, and how SSO policy can help avoid problems associated with patent assignments. Legal cases clarifying the potential issues are reviewed as well, with an emphasis on the United States. This discussion is in Chapter 5.
- Comment on the complex issues regarding the issuance of injunctive relief for FRAND-encumbered licensing arrangements of essential IPR. This area has become increasingly contentious, as discussed in Chapter 6.
- Relate the activities of SSOs to their interactions with government policy, such as cooperation with prior art searches in patent applications and disclosure and procurement policies. These issues are described in Chapter 7.
- Describe practices and draw lessons from the operations of SSOs in China, India and Brazil to understand the landscape and priorities of key emerging countries. This analysis is presented in Chapter 8.

The committee carefully deliberated whether to address issues surrounding injunctions, including exclusion orders issued by trade authorities such as the U.S. International Trade Commission, because they are not explicit in the statement of task. Members found it necessary to do so because of the inevitable and close relationships among determination and disclosure of SEPs, licensing conditions, and avenues for relief against infringement. A key question facing SSOs and regulators is determining under what circumstances a FRAND licensing commitment is incompatible with injunctive relief. The committee believed that failing to address this question would not fulfill the terms of its task.

Similarly, the committee decided early in its deliberations to discuss the question of whether and how national patent offices could fruitfully collaborate with SSOs in issues of prior art, affecting patent quality, and the recordation of patent transfers to enhance transparency in licensing SEPs. While this issue was not mentioned in the statement of task, the committee thought it an important element to discuss in the context of licensing transactions and one that could not be separated from the patent transfer question in any case.

The committee acknowledges that the issues involving patents in standard-setting play out in a broader context of national and international policy regarding intellectual property rights. These policies are in flux. For example, some member countries of the European Union are working toward a unified patenting regime and a single patent court. China, India, and Brazil have recently adopted significant changes in their laws that affect the scope of patents and, at least indirectly, the general conditions of licensing. In the United States, the patent system is changing in response to a variety of perceived problems – low patent quality, long delays in patent examination, excessive litigation and abusive tactics on the part of some patent holders.

The America Invents Act (AIA), passed and signed in 2011, ushered in the most significant changes in 60 years including enhanced post-grant opposition procedures, new opportunities for third party-submission of prior art, more resources for the U.S. Patent and Trademark Office, and adoption of the international norm for patent priority, first-inventor-to-file. At the same time, U.S. courts have limited the reach of exclusive patent rights in a variety of ways including ending nearly automatic access to injunctions against infringers,[4] raising the standard of non-obviousness,[5] and restricting patent eligible subject matter.[6]

Notwithstanding that these changes are still being implemented and their effects are uncertain, there continue to be calls for reforming U.S. patent law, notably to curb the opportunistic practices of so-called patent assertion entities (PAEs) or patent "trolls." These firms generally do not practice their patents and

---

[4] *eBay v. MercExchange*, L.L.C. 547 U.S. 388 (2006). This case is discussed further below.

[5] *KSR International. Co. v. Teleflex, Inc.* 550 U.S. 398 (2007).

[6] *Association for Molecular Pathology et al. v. Myriad Genetics, Inc. et al.* (Supreme Court no. 12-398, 2013).

often are not responsible for the inventions that lead to patents. Rather, they acquire patents for the purpose of threatening large and small practicing companies, which may or may not be infringing, with lawsuits in order to obtain royalties from licenses or monetary settlements or damage awards. According to a recent study by the Obama Administration, lawsuits brought by PAEs have tripled in the last two years and now constitute the majority of patent infringement suits (Executive Office of the President, 2013). Claiming that PAE activity disproportionately hurts small businesses and deters technological innovation in some sectors, the Administration supports legislation favored by a number of large operating companies to discourage such activity by permitting the courts to impose attorney fees on entities that bring unwarranted lawsuits. The Administration would also force patent owners and applicants to disclose the "real party in interest" to shed light on who benefits from PAE activity.

It is quite possible that such reforms, adopted and proposed, will or would reduce some of the pressures underlying the issues addressed in this report. For example, if an outcome of the America Invents Act and recent court decisions is to raise the quality of issued patents—i.e., increase the likelihood that they are truly novel and inventive—then litigation associated with the validity of claimed SEPs could be reduced. Transparency of patent ownership, especially in the case of patent transfers, would enhance the transparency of FRAND licensing commitments, as we discuss later in this report. However, while the committee recognizes the importance of a well-functioning patent system, proposing further fundamental reforms was not part of our charge from the U.S. Patent and Trademark Office or the National Research Council. In any case, it would require a committee with a quite different composition of expertise.

1.5 Economic Context

Issues of standardization and the associated management of intellectual property in technical standards have taken on ever greater importance in recent decades because of three major and interrelated factors. First is the rapid advance of globalization reflected in the growth of global production and innovation networks, both within and across enterprises (Ernst 2006, Maskus 2012). Second is the increasing integration of major emerging economies, such as China, India, Brazil, and Mexico, into world markets for goods, services, and technology. Third is rapid and even accelerating technological change in key sectors of competition where standardization and interoperability are critical elements of success, notably in the ICT sector. As a result, standards and policies need to evolve and adjust over time to reflect the changing technological environment.

All of these factors are transforming and deepening the challenges of effectively managing the development and use of SEPs in an environment of intensifying technological competition. Global enterprises increasingly see control over intellectual property, especially in essential architectural and interface standards, as the major determinant of competitive success. Indeed, control over IP offers one primary means of earning returns on increasingly costly research and development (R&D) investments. At the same time, it is important that implementers around the world have access to these standards to promote product-level competition and procure consumer benefits. Yet they may be developed by SSOs with different rules or implemented in countries subject to varying IPR and competition policies. These pressures explain the need for increasingly sophisticated approaches to managing SEPs in standardization.

Innovation is at the core of both national and enterprise-level strategies for competing in global markets in order to encourage growth and job creation (OECD, 2012). Intensive investments in research and development (R&D) were, until recently, largely confined to institutions and firms headquartered in a small number of advanced economies. Today, however, R&D expenditures exceed 1.5 percent of GDP (a previously high benchmark) in dozens of nations, including China, Korea, and Singapore, where the growth of such investments has been particularly rapid.[7] At the firm level, these increases have emerged both within national enterprises and through international investments in R&D performed by affiliates of multinational companies. The latter trend in particular demonstrates the globalization of innovation efforts through research networks (OECD, 2008). Distributing R&D among affiliates can reap several competitive advantages, including reduced personnel costs, enhanced intellectual diversity, greater access to local fiscal subsidies and markets, and higher sales of products and technologies developed for local or regional markets (Ernst, 2006).

Investments in R&D aim to produce new process technologies and higher-quality varieties of goods and services. A central component of this research is the development of technical standards, which exist and are often updated in virtually all industries. Standards often build on scientific knowledge from research in biology, informatics, mathematics, physics, and other basic sciences, explaining why universities and public research laboratories' such as the National Institute of Standards and Technology (NIST) in the United States, interact with standards organizations. Many technologies, such as those in software and electronic communications, emerge from research in engineering work at both grant-supported laboratories and private firms, requiring cooperation and collaboration among these institutions to achieve standardization.

In this context, two types of R&D investments are fundamental for the commercialization of technologies in several industries. First is the development of a new process technology or product itself, which is largely a private affair though it may be based on the outcomes of knowledge generated via publicly supported basic science. Second is the specification of standards that these technologies and products must meet in order to enter the marketplace.

Standards are developed by a complex mixture of private and public interests. They may be the result of individual firms discovering and promoting a workable specification that achieves customer acceptance. Indeed, standards

---

[7]Data are from World Bank, *World Development Indicators*. See http://data.worldbank.org/data-catalog/world-development-indicators.

development has become an important field of corporate strategy in some industries. Standards are also developed by SSOs, which consist of representatives of private and, often, quasi-public entities. They may also be set by government agencies, whether to address a clear market failure or to promote local industrial-development objectives. Governments may regard the standard-setting process as an integral part of national competitiveness and innovation strategies. More broadly, public authorities may be involved in standards development and diffusion because of the recognized benefits they entail. For example, early in 2013 the European Union put in place a new regulation addressing the benefits of standards under recognized rules.[8]

Within this complex framework firms and SSOs act collaboratively to enhance efficient standardization. Indeed, standards selection often involves difficult engineering decisions to achieve compatible specifications across a variety of complex technologies, while ensuring that any promulgated norms are compatible with user needs. Thus, especially in industries with multiple competing technologies it is necessary to have extensive upfront technical and management consultations to arrive at the most appropriate standards, the fundamental purpose of SSOs. Often the technologies entering standards are patented. Thus, central to this management of standards is the specification of guidelines and rules for the licensing of standard-essential patented technologies.

At the same time, there may be numerous opportunities for firms to engage in exclusionary practices or exploit market power. These problems are likely to be most significant where technologies entering a standard are patented and essential for use, innovation is cumulative, and there are important network effects and interoperability needs. The main competitive concerns of this type, arising from the combination of standardization and the assertion and enforcement of patents, are addressed in detail in this report. Through their IPR policies regarding disclosure, licensing, and transfers of patents, SSOs could, in principle, effectively diminish such problems. Where these policies are inadequate for this purpose, however, competition authorities and the courts play an important role. Thus, in this report the committee offers guidance for both SSOs and public agencies.

1.6 Standardization in the ICT Setting

For the past two decades the information and communications technology (ICT) sector has represented one of the most dynamic commercial markets and most active arenas for standards development. In particular, the convergence of products and services in the Internet and cellular communications sector has transformed the daily lives much of the world's population and spawned industries in every part of the world with aggregate economic activity approaching $2

---

[8]See http://eur-lex.europa.eu/LexUriServ/LexUriServ.do?uri=OJ:L:2012:316:0012:0033:EN.pdf.

trillion per year.[9] The Internet and cellular communications industries had separate origins and until recently followed separate development paths. However, they have converged rapidly in the past five years and it is likely that by now, the majority of devices accessing the Internet use wireless communications. Computers, tablets, phones, and televisions increasingly share these capabilities. With the ability to connect devices anywhere in the world to one another, the Internet and cellular networks rely heavily on interoperability standards developed by SSOs. These standards almost inevitably incorporate patented technologies.

The technologies embodied in today's complex microelectronic products, such as a smartphone, are governed by hundreds of standards, developed by many SSOs, each with its own culture, governing principles, and specific processes. Indeed, the heterogeneity of practices across SSOs is striking. The following paragraphs describe briefly four multinational SSOs with diverse characteristics.

- The Third Generation Partnership Project (3GPP) formulates standards for cellular communications. It is a consortium of six national and regional standards organizations, referred to as "organizational partners," in Europe, the United States, Japan, Korea, and China. Individual members, which are companies associated with one or more of the organizational partners of 3GPP, contribute to producing the standards. Partner SSO policies govern.[10]
- The Institute of Electrical and Electronics Engineers Standards Association (IEEE-SA) publishes standards in many categories. Several devices implementing IEEE standards for wireless local area networks (Wi-Fi) and Bluetooth communications are critical components of complex computers and smartphones. IEEE-SA has individual and corporate members and is part of the IEEE, a transnational professional association.
- An important source of standards governing Internet communications is the Internet Engineering Task Force (IETF). Its goal is "… to make the Internet work better by producing high quality, relevant technical documents that influence the way people design, use, and manage the Internet." Reflecting the organic, bottom-up evolution of the Internet, IETF activities are open to anyone and there is no formal membership or membership fee.

---

[9]Chetan Sharma estimated worldwide mobile service revenues to reach $1.5T in 2012 and iSuppli projected 2012 factory revenues for mobile communications equipment to reach $376 billion.

[10]Because there is no independent 3GPP IPR policy, this organization was not included in the survey described in Chapter 2.

Introduction

- The International Telecommunication Standardization Sector (ITU-T) of the International Telecommunication Union (ITU), another source of communications-based standards, is the United Nations specialized agency for standardization in information and communication technologies. The ITU currently has a membership of 193 countries and over 700 private-sector entities and academic institutions. Currently, one of the most prominent ITU-T standards, called "Recommendations," is the H.264 standard for video compression.

The rapid evolution of ICT technologies and the growth of ICT markets are fueled by intense activity in these and other SSOs described in this report and hundreds more. The work of each SSO is generally conducted by technical committees within the organization that are responsible for individual standards. For example, 3GPP has four Technical Specification Groups, further subdivided into 17 working groups. The IETF's standards development work is organized into eight areas, subdivided into more than 100 working groups.

Each SSO has a unique background, technical scope, and rules regarding membership and participation. Similarly, each SSO has its own set of procedures for introducing, adjusting, approving, publishing and revising candidate standards.[11] This diversity is also reflected in the way the SSOs treat intellectual property related to candidate standards and adopted standards. Each SSO (in the case of 3GPP, each of its organizational partners) has its own policy regarding disclosure of essential patents, licensing commitments, and in some cases, transfer of essential patents.

Although standards creation and revision is in large part an engineering discipline, the outcomes of the engineering efforts are strongly influenced, and in some cases mandated by laws, regulatory practices, and judicial systems of the jurisdictions of the SSOs and places where the standards are implemented. Until recently, the sources of standards and the markets for products and services that implement the standards have been concentrated in regions with highly developed economies. However, ICT markets in these regions are beginning to saturate and most of the near-term growth is likely to come from emerging markets, most prominently, China, India, and Brazil. Although they differ from one place to another, the laws, regulatory practices, and judicial approaches to standards are relatively well-established in economically developed countries. By contrast, these practices are still evolving in emerging markets, including those studied by the committee—China, India, and Brazil. Among these three countries, China has implemented the most extensive set of institutions and procedures and Brazil the least.

---

[11] As noted in Chapter 2, however, the large majority of standards bodies accredited by the American National Standards Institute (ANSI) simply adopt the ANSI IPR policy as their own, with little variation.

1.7 Stakeholders in Standard-Setting

There are a variety of interested actors involved in setting standards and managing and licensing intellectual property in ICT and other sectors, raising questions about finding the appropriate balance among them. The most obvious stakeholders are members of SSOs, including for the most part companies that manufacture products or market services that use the standards, companies that operate networks that practice the standards, firms that develop or acquire technologies purely for licensing, and in some cases academic institutions and government agencies. Many SSO members own patents they consider to be essential to implementation of the standards. Patent owners may have diverse motivations and strategies for utilizing their intellectual property. Some use these assets primarily for defensive purposes, such as to establish a form of détente or enable voluntary cross-licensing of patents among companies, supporting their "freedom to operate" in the marketplace. Others use their patents to generate a stream of royalty revenues; acquire patents to exercise the right to exclude others from using their technologies, as authorized by patent laws; or view patents as assets that help enable financing or entry into new markets where competitors have patent portfolios as well. Many firms have acquired patents for several, or perhaps all, of these and other reasons, following different business models.

There are also many entities that are not members of SSOs but are strongly influenced by the intellectual property policies of those organizations, including owners of essential patents, implementers of standards, and end users. These categories encompass individuals, companies, universities, and governments. For example, in the United States federal agencies are required to adopt private-sector standards under the National Technology Transfer Advancement Act, signed in 1996 (104 P.L. 113; 110 Stat. 775), and Circular A-119 of the Office of Management and Budget (OMB), revised in 1998 (OMB 63 FR 8546, 1998), whenever possible.

Further, it is becoming increasingly common for enterprises to acquire large portfolios of patents, many of which may be SEPs, to gain their implicit economic value. A few recent examples of companies that have obtained large bundles of patents from their original owners are Google, owner of patents acquired by purchasing Motorola Mobility, IPCom, purchaser of patents from Bosch, and the Rockstar Consortium, composed of Apple, Microsoft, EMC, Ericsson, BlackBerry, and Sony, that purchased Nortel's patents during that company's bankruptcy proceedings. Such acquisitions raise questions about the transferability of prior licensing commitments.

1.8 International and Multilateral Issues

The committee was asked to address geographical variations in practices and policies, with a view toward identifying their effectiveness in reducing conflict among stakeholders in this area. In this regard the report focuses on three salient features of the global system. First, as described in Chapter 2, there is considerable heterogeneity in practices across SSOs that are located in different major economies. Keep in mind, however, that the most prominent SSOs in ICT are trans-boundary institutions, with membership from multiple locations and uniform rules applied to those members. This sets up an interesting hierarchy of organizations among those that arguably are more national in scope and the major international SSOs, creating opportunities for both collaboration and discord.

Second, a meaningful distinction exists between the government policies of the established industrialized economies and those of many emerging countries. The mature industrialized countries—the United States, members of the European Union, Japan, and to a large extent South Korea—generally share a common policy environment that relies on private sector organizations to develop and implement technologies through decentralized market competition, although university and government research laboratories are important as well.

The situation in key emerging economies is different. Chinese authorities, for example, tend to view standards setting as a centralized, top-down process that may achieve a variety of objectives, including domestic industrial policy and inbound technology transfer. Thus, although Chinese representatives are active in many international SSOs, the government also emphasizes and guides the development of domestic standards in key technologies. In short, China approaches domestic standards-setting more as an element of public policy and management than the province of the country's enterprises. For their part, India and Brazil are only beginning to develop their standards policies in ICT and it is too soon to know which approach they may follow.

Third, the committee asked whether there are significant differences in laws and court opinions across jurisdictions regarding key elements of IP management in SSOs and firms. Such differences could arise, for example, in the areas of understanding the meaning of FRAND commitments, injunctive relief, transferability of licensing commitments, and public recording of patent transfers. In this context, the committee believes there is scope for greater communication and collaboration among SSOs and the patent offices of major economies, and among courts to reduce legal dissonance.

To the extent that differences in laws and public policies regarding SSOs pose significant difficulties for cross-border trade one might support efforts at international policy or law harmonization. This might be achieved through reformulating the Agreement on Technical Barriers to Trade (TBT) or the Agreement on Trade-Related Aspects of Intellectual Property Rights (TRIPS) at the World Trade Organization (WTO) or through additional undertakings on IP standards at the World Intellectual Property Organization (WIPO). The committee chose not to pursue this line of inquiry, believing that it is unclear that there are sufficient problems with the current international system to justify recommending multilateral negotiations that are difficult to initiate and conclude in either venue, and in any case would involve broader issues of trade policy outside the committee's statement of task (Maskus, 2012). Thus, in the near term we see little role for the WTO or WIPO in this context, though further study of the potential for such involvement in the longer term is warranted. In all likelihood, greater societal benefits are available from simply restricting or eliminating nationalistic policies and practices that unduly reduce competition and innovation, a role for national authorities.

2. A Comparison of SSO Policies and Practices

2.1 SSOs Surveyed for the Study

This chapter reviews and compares the primary approaches to policies regarding intellectual property rights management and licensing rules across 12 major standard-setting organizations operating in the information and communication technology space. This material is abstracted and summarized from a background paper written by two members of the committee available online (Bekkers and Updegrove, 2012).[1] In consultation with the full committee, the authors selected these organizations of various types of SSOs, including the "business models" they and their members tend to follow, inclusive of several geographic membership models (national, regional and global), and comprehensive in sectoral breadth within ICT. Given available time and resources it was not possible to survey more than a selection of the many hundreds of relevant SSOs in ICT, an aggregation that seemingly grows larger each week. The information gathered from documentation, questionnaires and telephone inquiries was current through 2012. The committee is aware that several organizations, including ETSI, ITU, W3C and IETF, are considering changes to their IPR policies. The SSOs surveyed are listed below.

> International Organization for Standardization (ISO)
> International Electrotechnical Commission (IEC)
> International Telecommunication Union (ITU)
> Institute of Electrical and Electronics Engineers (IEEE-SA)
> European Telecommunications Standards Institute (ETSI)
> American National Standards Institute (ANSI)
> Internet Engineering Task Force (IETF)
> Organization for the Advancement of Structured Information Standards (OASIS)
> VMEBus International Trade Association (VITA)
> World Wide Web Consortium (W3C)
> High Definition Multimedia Interface (HDMI) Forum Nearfield Communications (NFC) Forum

---

[1] See http://www.nap.edu/catalog.php?record_id=18510.

The first three SSOs are the International Organization for Standardization (ISO), International Electrotechnical Commission (IEC), and International Telecommunication Union (ITU), which is a United Nations treaty organization. These are large, formally recognized SSOs with global membership consisting of nationally designated private or public-sector representatives, although individuals and firms participate as well. These groups develop standards through extensive collaboration with national standards bodies. Notably, in 2007 they adopted a common patent policy and set of guidelines. While this decision resulted in a largely harmonized set of IPR rules, each of the three SSOs retains some flexibility for implementing specific requirements.

Next is the IEEE (originally the Institute of Electrical and Electronics Engineers), a professional association with over 300,000 individuals around the world as members and the publisher of numerous technical journals. Standards development occurs in the IEEE-Standards Association (IEEE-SA), which is responsible for several critical specifications, including Ethernet, Wi-Fi, and Firewire.

The fifth SSO is the European Telecommunications Standards Institute (ETSI), a body focused on information and telecommunications technologies, including fixed, mobile radio, converged broadcast and Internet technologies. Although founded as a regional group, some of its standards have become adopted around the world. For example, ETSI developed GSM, a highly successful standard for mobile telephony, and participated in construction of its successor, the 3G UMTS/W-CDMA specifications. ETSI has an extensive IPR policy, which evolves over time, sometimes under the influence of the European Commission. Since 2003, it has collaborated with the European Patent Office to expand the latter's database to include thousands of technical contributions made to ETSI as part of the standard process. These documents can now be searched for prior art.

The sixth entity is the American National Standards Institute (ANSI), which is not a standards-setting organization. Rather, it is a national, nongovernmental organization that supports standards development in the United States and the standards-related interests of America abroad.[2] As part of this role, ANSI accredits SSOs—approximately 200 in all—with respect to their standards development activities. If these SSOs fulfill certain criteria, called "essential requirements", they can create standards that ANSI will approve as "American National Standards," of which there are now over 10,500. ANSI includes adherence to its IPR policy among these criteria, defining a baseline set of rules that accredited organizations must meet, though these groups have leeway in setting their own procedures so long as they are not in conflict with the essential requirements. For purposes of this report, we will refer whenever possible to the ANSI's IPR requirements for accredited SSOs as if they were those of an actual SSO. In fact, a large majority of its accredited SSOs follow the ANSI IPR policy without variation.

---

[2]Many international companies with U.S. business operations are members of ANSI.

A Comparison of SSO Policies and Practices 33

Next is the Internet Engineering Task Force (IETF), a group of academic, industrial, and government engineers working to improve the technologies driving the internet. Among its other functions, IETF develops Internet standards, the most famous of which is the TCP/IP protocol suite of programs. There is no membership in IETF per se and standards are adopted on the basis of a "rough consensus." Regarding patents, IETF prefers to adopt standards not encumbered by known IPR claims, unless protected technologies offer considerable technical superiority to available alternatives. Patents and patent applications covering technologies under consideration by IETF are required to be disclosed as early as possible in developing technical standards.

The eighth group is the Organization for the Advancement of Structured Information Standards (OASIS), which focuses on e-business and web service standards. OASIS is often referred to as a consortium for it operates outside the typical infrastructure of traditional SSOs. Through a series of revisions of its IPR policy, OASIS has developed a new and more flexible approach to assigning obligations relating to essential claims in patents. In particular, in 2005 it adopted a "multi-mode" IPR regime permitting a working group to operate under a set of guidelines that either would or would not allow participants to require payment of licensing fees under a FRAND arrangement. This approach has now expanded to four modes, including one based on a non-assertion of patents.

The ninth organization is VITA (originally VMEBus International Trade Association), which works on standardization in a variety of electronics areas, such as avionics and other military and industrial applications. This SSO is the only one in this set that requires members to make *ex ante* disclosure of the most restrictive licensing terms, including economic terms that the SEP owner reserves the right to demand. VITA is also unusual in that it declares its intention to enforce all disclosure terms and conditions on essential claims, take action against frivolous assertion of SEPs, and require members to submit to binding arbitration when conflicts arise.

The next SSO is the World Wide Web Consortium (W3C), which develops standards used in connection with the Web and other technologies. Consistent with the development culture of the Web, the W3C in 2003 adopted a license-fee-intolerant patent policy. Its members see the Web as basic global infrastructure that needs the widest possible distribution of its technologies and standards. Thus, it pursues a FRAND royalty-free policy.

Our eleventh SSO is the High Definition Multimedia Interface (HDMI) Forum, a consortium which focuses on standards for a compact interface among compliant devices to share uncompressed digital audio/visual data. Implementers of the proprietary HDMI technology must pay royalties to developers of patented technologies in the standard, including the original seven founding companies. Further developments of the standard from its initial form are subject to agreements under which patent owners agree not to assert any essential claims against implementers.

Finally, the Nearfield Communications (NFC) Forum is a consortium founded in 2003 to develop and market its short-range wireless interaction standard enabling data exchanges among consumer devices, for example, to facilitate payment transactions. All members must respond to requests for disclosure and licensing of SEPs to reduce the possibility that such claims will be asserted later to the detriment of an implementer.

To summarize, four of the organizations considered here are formal SSOs, one is a standards accreditation group, and the remainder are consortia of companies and/or individuals. All have global membership, though ETSI is largely European-based and IEEE-SA and ANSI are largely U.S.-based, and the standards activities and IPR rules of all the SSOs have international reach. Some work in broad areas of technology, while others focus on specific technical specifications. In the aggregate they span the ICT space, including consumer electronics, telecommunications, the Internet and Web, and related areas.

For purposes of developing the materials summarized below, the authors of the background paper collected available documents from each SSO that are relevant to organizational IPR policies. These policies were analyzed in a standard format drawn largely from the American Bar Association's Standard Development Patent Policy Manual (2007). The authors further developed a series of questions posed to SSO representatives, along with an invitation to comment on the accuracy of their analysis. Consistent with the basic charge to our committee, the findings in the background paper, and those summarized in this chapter, are descriptive in nature rather than assessments of effectiveness, which in any event would depend on a large set of economic, technological and social factors for each SSO. Later chapters of this report offer deeper assessments of some key standards-related IPR policy questions.

2.2 A Note on Terminology

Readers of this report who are already familiar with standards and intellectual property management know that there are many intricacies and complexities in how policies are conceptualized and put into operation in licensing markets. This means, among other things, that there may be multiple names for closely related concepts that, while they may vary importantly in legal contexts, can be captured with a single overarching title. Doing so permits us to discuss principles without having to cover all possible outcomes, except where we need to be precise to avoid confusion. This approach should clarify our analysis throughout.

In that context, we will use the following blanket terms, recognizing there are differences among elements covered.

- **Standards** is our word referring to numerous similar terms, including standards, specifications, and recommendations.

A Comparison of SSO Policies and Practices 35

- **Standards-setting organization (SSO)** incorporates all variants of groups that develop standards, including Special Interest Groups (SIGs), standards-development organizations (SDOs), consortia, and other entities. The acronym SSO is often used interchangeably with SDO but, in principle, the former term covers the activities of both setting and managing standards, including associated intellectual property issues. Hence, we opt here for SSO.
- **Fair, reasonable and non-discriminatory (FRAND)** is used to cover both RAND and FRAND commitments, whether royalty-free or otherwise. FRAND is more commonly used in Europe and RAND in the United States but the words are sometimes used interchangeably. Our choice of the acronym FRAND is meant to cover both terms.
- **Standard-essential patent (SEP)** refers to both patents that are essential to the use of a standard and the essential claims of such patents. Some IPR policies focus on essential patent claims rather than SEPs and, indeed, a properly crafted rule rarely imposes licensing obligations to claims other than the essential ones. However, the analytical concepts involved are largely common to both terms and SEP appears to be in wider use in the literature. Note also that the word "necessary" is sometimes used in place of "essential."
- **Licensing commitments** and **licensing obligations** are used interchangeably to cover a wide set of activities, including letters of assurance, declarations of licensing positions, licensing statements, licensing undertakings, and similar pledges.
- **Disclosure** refers to the various processes within which a firm or individual informs an SSO and other entities that it owns or is aware of a patent, or patent application, that may be relevant for a standard. This usage is different from the concept of disclosure at a patent office, which refers to the information an applicant must provide to satisfy certain requirements for obtaining a patent.

2.3 A Caveat on Coverage

We note at the outset that the forms of intellectual property to which we refer in this report are limited to patents, whether essential to a standard or otherwise. Other IPRs certainly arise in this area. For example, SSOs or other entities may own copyrights in a written technical standard or manual, but typically this is relevant only to the reproduction and distribution of those materials and not to the implementation of standards. More relevant is computer software that must be deployed as a necessary component of a technical standard, making it a form of "essential copyright." In most cases, such software is used for such purposes as defining required outputs, as opposed to being used by an implementer in its product. As the background paper notes, such essential copyrights are only addressed by a small number of SSOs.[3] Copyrights are an area in some flux regarding ICT standards and are likely to become more important over time.

Firms employing or developing standards may own trademarks and similar distinguishing characteristics, which may implicate some licensing considerations in the marketplace. The ITU-T has a set of guidelines in this area and ANSI also provides related guidance. Some SSOs in the ICT area pay relatively little attention to trademark issues. For example, the NFC Forum clarifies that contributions of rights to trademarks are not required under its policy. Nor are trademarks included in the IETF definition of essential IPR. ANSI generally discourages the required use of trademarks in developed standards. Nevertheless, trademarks and certification marks play an important role in certifying that different products are compliant with specific standards. Compliance certification and trademark usage are not addressed in this report.

Other forms of intellectual property that are sometimes mentioned include utility models, inventor's certificates, and database rights. It seems that IPR other than patents rarely have been disclosed under terms of SSO policies.[4] While not directly relevant to ICT, plant patents and plant variety rights may take on increasing importance in standardization in other fields, such as biotechnology and life sciences. Finally, many SSOs do not expect members to disclose trade secrets because their disclosure likely would cause the trade secrets to lose confidentiality. In general, where trade secrets are revealed as part of a license the parties must arrange contractual terms to sustain secrecy.

In the committee's view, therefore, IPR elements other than patents are secondary to the issue of essentiality when it comes to the licensing and access procedures that are the subject of this report.

2.4 SSO Approaches to Basic IPR Issues

Given the considerable complexities involved in the interplay between standards development and IPR licensing, both in themselves highly complicated subjects, it is impossible to review all of the variations in SSO policies in this overview section. Rather, we take a high-level view of the essential principles involved in order to set the foundation for later discussion in the report. Still, we are keenly aware that details matter and that any summary of this kind may leave open as many questions as it answers. Further, we are aware that SSO policies may be changed at any time. Thus, we encourage readers interested in spe-

---

[3] Of the 12 SSOs surveyed, the IPR policies of six include essential copyrights and those of the others do not. IETF includes them in essential IPR and treats them in the same manner as patents. Bekkers and Updegrove, op. cit., p. 36.

[4] Design patents that cover new and ornamental designs (as opposed to utility patents that cover useful inventions put to practical applications) are typically not involved in standards because alternative designs may be employed to make patented versions non-essential. However, the rapid increase in design patents to protect computer-generated imagery, such as graphical interfaces, likely will lead to their growing role in standards.

cific policies of particular SSOs to consult the background paper, from which this summary is drawn.

With respect to prior literature, to the committee's knowledge no one has made a comprehensive investigation of SSO patent policy formulation. Lemley (2002) took a first step by documenting substantial heterogeneity in IPR policies across SSOs. This chapter complements his study by taking a more detailed look at the policies of a smaller number of organizations. We find substantial variation in aspects of IPR policies that were not examined by Lemley, such as the definition of essentiality, rules for disclosure of third party patents, the mechanisms for establishing licensing commitments, and the scope and revocability of those commitments.

As we note below, SSOs rarely state their objectives. Thus, it is hard to determine what motivates differences in IPR rules, or to evaluate how SSOs perform in relation to stated goals. Lerner and Tirole (2007) assume that policies are chosen to benefit members (modeled as a single technology sponsor), subject to the constraint that later adopters must find the SSOs' "certification" sufficiently credible. In our view, the Lerner and Tirole approach captures some important features of the policy-formulation process. However, it would be equally valid to model SSO policy formulation as an open-ended negotiation among prospective members with varied interests, often building on an existing set of rules and policies that forms an important reference point in such negotiations. Farrell and Simcoe (2012) discuss the role of inter-SSO competition in this process, observing that such competition may take different forms with correspondingly different effects for competition and consumer welfare.

Ultimately, the committee did not find sufficient systematic evidence to take a strong stand on how the SSO policy formulation process works, or might be made to work better. However, there was a consensus among committee members that encouraging SSOs to be more explicit about their policy goals might help those who wish to build the empirical foundations for a better understanding of the rule-setting process. Despite this lack of clarity regarding processes within SSOs, the committee was able to arrive at consensus recommendations regarding a number of specific policies.

Articulating policy goals

It may be surprising that few SSOs state explicit goals for their IPR policies, despite their evident importance. There are many objectives that SSOs could pursue, even if unstated, ranging from promoting widespread adoption of their standards with minimum restraints on access arising from IPR and ensuring that each essential IPR is available on reasonable and non-discriminatory terms, to ensuring fair compensation for SEP owners and providing enough structure that good-faith licensors and licensees understand their rights and obligations. The general absence of stated objectives presumably reflects the difficulty of reaching agreement on organizational aims among SSO members with disparate interests. Companies with large patent portfolios that are active in multiple SSOs may see different opportunities and burdens in specific policies than do smaller participants with few patents. Further, SSOs with broad coverage across technology classes may find it difficult to determine a set of objectives that make sense throughout their activities. Organizations focused on very specific technologies, such as W3C, HDMI Forum and NFC Forum, may find it easier to articulate their underlying goals.

One potential concern about the general absence of articulated objectives is that SSOs may find it hard to evaluate their IPR policies. For example, while some SSOs generally posit the importance of essential IPR disclosure for selecting and disseminating their standards, it may not be that the actual disclosure processes are effective for that purpose. A related concern is that organizations rarely attempt to define critical concepts and steps, such as FRAND terms and conditions or disclosure timing.

Defining essentiality of patents and claims

In general, a SEP is a patent protecting an invention required to practice a given industry standard, so that infringing essential patent claims is unavoidable when implementing the standard. Defining essentiality is important both for disclosure and licensing obligations within SSOs.

In their IPR policies, all 12 SSOs address the need to declare possibly essential patents or the possible ownership of SEPs. However, just six include copyrights essential to implementation of a standard within their definitions. In these cases the essential copyrights are treated under the same language as SEPs, which may be problematic in light of key legal differences between patents and copyrights. However, ETSI has a separate IPR policy for copyrights including an explicit software license. IETF requires that any software source code (essential or non-essential) included in a standard must be available under an open-source license.

While some SSOs define essentiality to mean there are no technological alternatives, others also mention the concept of "commercial essentiality," meaning that non-infringing alternatives may exist but are too expensive or cumbersome to be worth bringing to the market. This situation is particularly likely in ICT because once a standard is widely adopted it may become economically infeasible to deploy an alternative workable technology due to network economies. In fact, just two SSOs (IEEE and VITA) include commercially essential patents within their IPR policies regarding disclosure and licensing commitments and one (ETSI) explicitly rules the concept out. To be sure, whether a technology is commercially essential is often a subjective issue, which may explain why some SSOs do not consider that notion in their IPR policies.

A further distinction of note is that within a SEP some claims may be considered mandatory for successful implementation of a standard. Others, however, may read on optional portions of the standard but could be quite important if the implementer chooses to conform to that optional set of requirements and are therefore referenced in the standard.[5] Licensees may wish to incorporate both within a transaction, while some licensors may prefer to withhold the optional portions or negotiate separate arrangements with terms and conditions that may not meet FRAND guidelines. In this case the implementer may find it difficult to avoid infringement when it sees the need to use the optional claims or patents. This situation raises the question of whether such optional claims should be included within the ambit of SSO IPR policies. Our inquiry revealed that the policies of the three large, global SSOs—ITU, ISO, and IEC—restrict their definitions to technically mandatory claims, which may or may not include required patents that read on the optional portions of the standard in question. Two groups—IETF and VITA—do not define optional claims, suggesting they are subject to licensing rules. ANSI rules leave it up to the SSO to decide whether optional claims are or are not subject to licensing obligation. The other six SSOs formally include optional components of SEPs in their IPR regimes.

We mention one other important element on which the surveyed SSOs have widespread consensus. Ten of the twelve clarify that essentiality is defined with respect to patents necessary for implementing the final standard. Several technologies may be disclosed as essential during development of the standard but not all of them may turn out to be necessary for the final specification. Only those that are ultimately necessary retain essentiality for purposes of the licensing obligations in 10 cases.

Despite the attention of SSOs to the essentiality issue at the broad level, Bekkers and Updegrove find a surprisingly wide variation as well as imprecision regarding how this extremely important element should be defined. In a number of cases there is a neglect to address important concepts or there was a decision to use vague language. We note that this outcome may result from the negotiations leading to formation of the SSO or adoption of its IPR policy.

Disclosure of SEPs

Most SSO IPR policies have two core elements: (1) rules regarding disclosure of patents that may have essential claims and (2) rules regarding licensing commitments or statements of non-commitment. While conceptually distinct, disclosure and licensing commitments are often intertwined. For example, patents are often disclosed in the same declaration form that includes licensing commitments. These relationships vary in complex ways across organizations. For example, in some SSOs a respondent who fills out a licensing statement signals that it believes it owns patents that will likely become SEPs, which must be disclosed. In others, participants may reserve the right to seek a paid license but are allowed to submit blanket disclosures that do list specific patents that may not be essential.

---

[5]These may be distinguished from non-essential claims within a patent. The policies of 11 SSOs exclude non-essential claims from the definition of essentiality and, therefore, from associated IPR rules.

This lack of a disclosure obligation generally arises in broader and more comprehensive cross-sectoral SSOs because its members may be large enterprises with very large patent portfolios. Determining which patents may be essential can be costly and burdensome. Making a FRAND commitment for any SEPs that the patent holder ends up holding regarding the final standard ensures that it will not use any of its SEPs to block implementation of the standard. It also reserves the right to license for a fee, which keeps the patent owner's options open while avoiding the need to scrutinize its patents. Indeed, most IPR policies do not mandate patent searches because of the cost implications. Thus, the scope of disclosure obligations reflects a balance between the benefits of providing substantial information and the burdens of providing it. This balance means that disclosure is effectively less than it might be with complete information revelation.

Most of the policies of the surveyed SSOs seek formal disclosure at some point in the standardization process and it is typically tied to an expectation that the member concurrently state its intentions regarding whether it will license its SEPs under terms of the organization's IPR policies.[6] This obligation is intended to result in a legally binding commitment with respect to SEPs. However, since it may be made late in the standard development process, the need to avoid infringement of a SEP that is eventually not made available for licensing can result in a significant loss of time before a standard can be adopted.

To be more precise, all of the surveyed SSOs except the HDMI Forum, which has no formal disclosure mechanism, obligate those who submit a technology they wish to include in the standard either to disclose the specific patents they own and believe would be essential, or at least to indicate that they likely hold SEPs. ANSI leaves this matter up to the SSOs it accredits. Some require the submission to be on a royalty-free basis while others permit the submitter to reserve the right to charge a fee. Submitters are sometimes not permitted to refuse to license any SEPs in their own submissions, but they can choose to not license SEPs where they did not contribute their technology to the SSO in question. Similar options face participants in standards working groups, which are obligated to disclose relevant essential IPR. Two SSOs, ETSI and IETF, further obligate members who are non-participants in working groups to disclose SEPs, while this is encouraged but voluntary in ITU, ISO, IEC, IEEE, ANSI, and OASIS. The umbrella organizations ISO and IEC go further to obligate disclosure by those who receive a draft standard and have patents that may be essential for its use.[7] IETF and W3C impose a similar requirement and OASIS requests such disclosure, presumably because non-members may participate in drafting sessions.

---

[6]An earlier and informal process is the patent call, by which participants in a technical discussion are expected to reveal patents of which they are aware that may contain essential claims with respect to the ultimate standard.

[7]Receipt of a draft standard may occur, for example, through involvement in the process through a national standards body.

Finally, all of the SSOs have rules covering participants' disclosure to members of known or suspected SEPs owned by members, participants, or unrelated groups. Such disclosure is mandatory for working group participants in ITU, ISO, IEC, ETSI, OASIS, and W3C, while again ANSI leaves the matter to its accredited SSOs. It is encouraged by IEEE and IETF, while VITA requires disclosure if the member or participant is a licensee of a third-party patent that may be essential to the standard. VITA, W3C, and VITA waive this requirement if disclosure, even to the SSO's members, would violate a promise of confidentiality.

Note that these organizations vary considerably in terms of what constitutes knowledge of potentially essential IPR, subject to disclosure, on the part of members and participants. It could be undefined or extend to personal knowledge (IEEE and IETF), elements that could be discovered by reasonable efforts (ETSI), or even patents that could be found by a "good faith and reasonable inquiry" (VITA). Still, none of these SSOs requires patent searches, so the obligation to disclose typically has some limitations.

A risk of the emphasis on disclosure of specific patents that potentially could be essential to a standard—a policy aimed at minimizing the chances of missing relevant SEPs—is over-disclosure of IP by participants who reveal more SEPs and other patents than ultimately necessary to implement the final version of the standard.

Such problems could be reduced, and the list of clearly essential patents clarified, by combining formal patent searches with efforts to assess essentiality after the standard is defined. However, it is important to note that none of the surveyed SSOs requires a member to engage in a patent search for purposes of disclosure or has a formal process for adjudicating the essentiality of patents. This reflects concerns expressed by members, especially large companies with multiple technologies, that a formal search across their large portfolios of patents would be expensive and may not even be definitive in terms of unearthing relevant SEPs. For those patent holders who do not proactively seek FRAND licenses from implementers this cost is not defrayed by licensing revenues.

Among our sample, ETSI has the broadest disclosure obligation, because it applies to all members and all standards activities and prohibits the use of blanket disclosures. While the scope of the disclosure rule depends on the knowledge of individuals and the companies they work for, which may be limited if they are not working on the draft standard, the obligation in ETSI is nevertheless the widest of any organization in this study.

A contrasting case is W3C, which has a limited disclosure obligation. The reason is that members of this SSO commit to license SEPs on a royalty-free (RF) basis if they are participating in the development of the relevant standard, making disclosure necessary only when a member seeks to exclude a patent from the default RF commitment. This RF "default mode" is thought to encourage implementation of W3C's standards and is consistent with the organization's preference for access to technologies.

Timing of disclosure

SSOs must decide whether their IPR policies recommend formal disclosure early or late in the standardization process, which is a difficult tradeoff to be managed. For technical committees undertaking this approach, early disclosures of potential SEPs and patent applications offer the advantage of developing standards that choose among and combine these technologies, while bypassing less promising ones. However, it is quite difficult at the immature stage of a standard for members to determine which patents may be essential, reducing disclosure quality. Further, this risks selecting technologies that are outdated by the time the standard is released, while patent applications may be denied or have their claims significantly limited. In short, there is a high likelihood of inadequate information with early disclosure. Late disclosures help address this problem but raise the risk that a draft standard will have developed so far that it may be difficult to work around an as-yet undisclosed SEP. This could pose a significant access problem for implementers if the late discloser chooses not to make a FRAND licensing commitment.

This is a hard problem and its best resolution probably varies with individual standards and technology areas. In this context, it is not surprising that SSOs have widely varying, and at times vague, policies. The common ITU/ISO/IEC policy calls for disclosure as early as possible but this seems to mean at a stage where the standard is sufficiently mature that possible essentiality may reasonably be determined. ETSI requires disclosure in a timely fashion and has the ability to sanction members that engage in intentional delays. ANSI encourages early disclosure but does not spell out what this might entail. W3C's approach is similar. VITA goes furthest, with a policy specifying precisely when a disclosure must be made and extensive guidelines on timing of disclosure.

It is noteworthy that the majority of SSOs do not specify any procedure for updating information about essential IPR, such as the denial of a patent application or the expiration or legal cancellation of a patent, although the SSOs with data-sharing arrangements with the European Patent Office (EPO) are able to track developments in a patent application and in patent families.

Blanket or specific disclosures

A second dimension is the precision with which SEPs must be disclosed. The basic issue is whether disclosures should identify the specific patents believed to be essential or blanket statements that a company believes it owns likely essential patents without specifically identifying them. The former policy obviously provides more complete information to standards developers and potential implementers than does the latter and for that reason is preferable in principle. However, identification of specific patents may not always provide sufficient value to justify the cost of doing so. Sometimes there may be a preference for a company to make a FRAND commitment for any SEPs it has that end up being essential to the final version of the standard, as opposed to only making a

A Comparison of SSO Policies and Practices

commitment that runs to specific patents. This approach may be sufficient for effective implementation of the standard, especially where the patent holder largely uses its SEPs for defensive purposes.

Some SSOs—ETSI, OASIS, and VITA—require specific patent disclosures and do not permit blanket declaration, though the disclosures are often based on a trigger, such as the personal knowledge of the individual participating in a standard's development. The policies of IEEE and ITU/ISO/IEC allow parties to file blanket disclosures, though the ITU requires disclosure of specific patents if the owner is not willing to license them on a FRAND or FRAND-RF basis. At IETF, blanket disclosures are permitted only if the owner commits to license its patents on a FRAND-RF basis.

Disclosing patent families

A particular technology may be the subject of patent applications or grants in more than one country and the aggregation of these patents may be called a family. Many standards, especially in the ITC area, are implemented in numerous countries. Thus, it is important for firms to disclose all the jurisdictions in which it seeks or has protection for its SEPs, preferably as a family rather than as individual patents. The common policy of ITU/ISO/IEC states an expectation that families will be disclosed and ANSI recommends this. ETSI's policy specifies that its disclosure requirements are satisfied if at least one member of a patent family is disclosed, information about potentially related patents is generated by linkage to the EPO database. In practice, many ETSI members disclose more than one patent per family. The publicly available ETSI database now contains information on patent families due to a recent program of cooperation between ETSI and the EPO.[8]

Common disclosure templates and public release

The increasing prevalence of SEPs, especially in ITC standards, has raised the importance of precisely stated disclosure requirements and licensing commitments that share considerable commonality. As a result, SSOs increasingly require the use of standard forms to ensure that disclosed information is complete, clear, uniform, and easy to consult once published. In most cases, these forms are used both to make disclosures and to choose among available licensing options. Bekkers and Updegrove revealed that the majority of SSOs now use disclosure and licensing templates.

Nearly all the SSOs publish their formal disclosure documents, usually on websites. Again, ETSI is notable because its cooperation program with the EPO to develop the Database Restructuring (DARE) project matched disclosed pa-

---

[8]We note that the patent holder is not the entity providing this family information, which may raise questions about its accuracy in some instances. See Chapter 7 of this report for a review of this program.

tents with the EPO's patent database. Patents in the ETSI database are now considerably easier to access and compare across countries.

A related element of transparency is the extent to which the minutes and reports of working group meetings are made public, which might be of particular importance to patent examiners in determining prior art and to licensors and potentially courts seeking to determine whether alternatives to patented technology were available at the time of standardization. Most of the broader and longer established SSOs, including ITU/ISO/IEC, ETSI, and IETF, do release such documents and invite non-members to technical meetings where sensible. Narrower and more recent SSOs, primarily consortia such as VITA, the HDMI Forum, and the NFC Forum, maintain the confidentiality of their documents.

Licensing commitments[9]

In general, SSOs aim to ensure that licenses for SEPs are available to all implementers, or that owners will not assert their essential IPR against firms that develop standards-compliant products. The minimum objective for virtually all SSOs is to ensure that all known SEPs are available under FRAND licensing terms (10 of the SSOs), with some favoring or requiring royalty-free FRAND terms (six SSOs). If an SSO discovers that an essential patent is not available it will typically attempt to obtain such a commitment or 'design around' that patent. This section reviews the primary processes for promoting licensing.

There are at least four basic mechanisms for establishing licensing commitments. First, several SSOs—ITU/IEC, ETSI, VITA, and NFC Forum—set out a general obligation for its members to submit a licensing declaration, often triggered at the time of disclosure. Participants in this arrangement have the option of not licensing on FRAND terms. Second, ITU/ISO/IEC, and IEEE solicit declarations from members regarding patents they believe might be essential. These declarations are made at the time of receipt of such requests or promptly after; again one option is not to license on FRAND terms. Members may also take the initiative and disclose their own patents by submitting a declaration without waiting for such a request. Third, OASIS, W3C, and the HDMI Forum follow a "default model" in which the obligation to offer licenses arises from membership or participation, meaning that a licensing commitment is agreed to upfront. Even here however, there generally are certain opt-out provisions for firms unwilling to meet required licensing conditions, especially if they did not contribute the implicated SEPs to the process. Finally, at IETF, which has no formal licensing requirement, many participants voluntarily submit licensing declarations with their patent disclosures. A large percentage of these declarations indicate that patent holders are willing to license such patents on a royalty-free basis.

---

[9]For analysis of additional issues, such as package licensing and grant-backs of non-SEPs, see Chapters 3 and 5. The IPR policies of SSOs generally do not address these issues.

The IPR policies of several SSOs, including ETSI, ITU and ANSI, encourage patent holders voluntarily to make an early (i.e., well before the standard is finalized) statement indicating that they may own patent claims that could prove to be essential and clarifying whether they would be willing to license these claims under FRAND terms. Once SEPs are actually identified later in the process a separate formal declaration, whether specific or blanket, may need to be issued in some SSOs. The intent of such early assurances is to avoid a situation in which a SEP owner ultimately decides not to license and to make parties to the standard more comfortable about including those SEPs with declared commitments. Other SSOs policies do not mention this possibility.

Scope and revocability of commitments

In 11 of the 12 SSOs, a commitment to license under whatever terms applies to any and all implementers, underlining the basic interest of SSOs in disseminating their standards into wide use. One exception is the HDMI Forum, where the commitment extends only to the members of this SSO or those who have signed a licensing agreement with it. It is important to note that all SSOs state that licensing commitments extend only to use necessary to implement a particular standard and produce compliant goods. Other uses would not comply and therefore would not receive the protection of SSO-facilitated license commitments.

As to geographical scope, 10 SSOs specify that licensing commitments hold on a worldwide basis, while the other two (ANSI and IETF) do not mention a geographic context. Global commitments are presumably consistent with the non-discriminatory element of FRAND and assist in spreading standards. However, members of our committee are aware of cases where patent owners refused to license in certain locations on the ground that the policies of the relevant SSO did not apply there.

What does a licensing commitment actually cover? In five of the SSOs a commitment must include any SEPs that read on the final version of specific standard in question, regardless of whether these patents were actually disclosed by their owner. In two policies the commitment covers only patents actually disclosed. The ITU/ISO/ICU and IEEE policies permit submitters to choose between these models. OASIS makes a distinction between licensing commitments of participants, which must apply to all patents with essential claims under the standard, and non-participating contributors, where the commitment applies only to SEPs under their contributions.

The great majority of SSOs in the sample specify that licensing commitments are irrevocable, although a few of them permit "upgrades" under which terms may be replaced later by conditions that are objectively more favorable to licensees. Some organizations, including ITU/ISO/IEC, ETSI, OASIS, VITA, W3C, and NFC Forum, specify that specific licensing agreements are also irrevocable before expiration in order to prevent their cancellation against the will of good-faith licensees for other than cause.[10] They also permit exceptions in cases where the licensor is sued for infringement by its licensee on the latter's patents.

Virtually all SSOs make licensing commitments public. Consistent with its policy on disclosure, VITA informs its members of such commitments, which are also made available to implementers on request.

Reciprocity conditions

IPR holders that commit themselves to FRAND or other licensing conditions may be concerned that they could face a situation where they are obliged to grant a license to a firm that refuses to license its own essential IPR on the same standard or group of standards back under similar conditions. To prevent this, SSO policies allow firms generally to include a condition of reciprocity in the licenses they grant. There are two relevant forms of reciprocity. First, bilateral reciprocity means that the licensee must offer its own essential claims under the same standard on the same conditions (e.g. FRAND or FRAND-RF) to the licensor, but not necessarily to other members or implementers.[11] Second, universal reciprocity means that the licensee must also offers its essential IPR for the same standard on the same conditions to all implementers.

Again, there are differences in approach across SSOs. The policies of ITU/ISO/IEC, ETSI, and VITA allow bilateral reciprocity. This form is automatically the basis of licensing agreements under the HDMI Forum and the NFC Forum. OASIS and W3C permit universal reciprocity. Under the NFC Forum and ITU/ISO/IEC, licensors who have otherwise committed to royalty-free FRAND licensing can charge royalties to licensees who seek payment from them for SEPs in the same standard, a situation that is not considered a violation of non-discrimination. Finally, nothing is specified by IEEE and ANSI but its members often seek reciprocity of some kind.

Ex ante *most restrictive terms*

A patent owner, at an early stage in the standards development process, may make binding commitments regarding the maximum royalty fee or other licensing terms it will seek in licensing contracts. In principle, such information can help inform SSO decisions on whether to include the patented technology in the standard. It may also create an incentive for IPR holders to limit their royalty demands, knowing that a lower price could increase the likelihood that their technology will be included in the standard. Another benefit is that it may help implementers in their licensing negotiations, as an upper bound is known. Whether these effects in fact are common among members of SSOs is unclear,

---

[10]Interestingly, OASIS, VITA, and W3C require licenses to be perpetual, a stipulation that makes sense for W3C, where licenses are royalty-free.

[11]U.S. regulators are currently urging SSOs to clarify the specific types of reciprocity they regard as consistent with a FRAND commitment.

A Comparison of SSO Policies and Practices

while the net benefits of such disclosure are much debated. Just one (VITA) of the SSOs in our inquiry specifies a requirement for *ex ante* disclosure of licensing terms. It is voluntary under IEEE, IETF, and ETSI. Moreover, ETSI has a repository for its *ex ante* declarations although none has yet been made. In any event, patent owners are free to make statements about licensing terms outside the SSO setting.

Injunctive relief

None of the policies of the surveyed SSOs imposes any restrictions on what legal remedies a member or third-party beneficiary of a licensing commitment may pursue in court. We mention it here because such remedies have recently become a matter of considerable concern in courts and among regulators.[12] Some analysts argue that when a FRAND commitment exists only the economic terms of a license remain the subject for a legal dispute. This is significant because under the laws of some jurisdictions an injunction against the sale of goods will be limited or not granted by a court if the party alleging infringement can be adequately compensated by a monetary award. Others argue, however, that injunctive relief is a key remedy an IPR owner should be entitled to seek if its patents are infringed.

2.5 Transfers of Licensing Commitments

An important subject for our committee is the transfer of patents with essential claims subject to licensing or non-assertion obligations. This has become an increasingly key issue with the proliferation of sales of patent portfolios and legal variations across countries about the treatment of such commitments in the event a patent owner goes bankrupt. The basic question is whether licensing obligations travel along when SEPs change ownership: is the new owner bound to the same commitments? Further, how far do the commitments travel in the event of successive ownership changes? Another question is the extent to which an original patent owner may incur any liability to someone sued by the transferee for infringement, if that owner had not informed the recipient of an encumbered licensing commitment.

In this section we review the approaches taken by our surveyed SSOs.[13] In summary terms, five SSO policies—ITU/ISO/IEC, ETSI, IEEE, VITA, and HDMI Forum—state that obligations must be transferred with patent ownership, thereby requiring that the patent holder bind its successor-in-interest. OASIS mentions the issue only in the context of bankruptcy. ANSI and IETF do not specify a policy.

More specifically, the ITU/ISO/IEC common policy has a strong section on patent transfer, defining obligations for the original patent holder. Essential-

---

[12] The committee considers this question in detail in Chapter 6.
[13] This transfer issue is discussed more fully in Chapter 5.

ly, a patent holder that has entered into a licensing commitment must ensure that the transferee also is bound to the same commitment. The clear intention of ITU/ISO/IEC is to have a strong set of rules on transfer of obligations, but there remain some ambiguities. For one thing, it is not clear if this policy applies only to participants in the drafting of a standard or also to all parties that submit a licensing declaration, although the latter are likely included. Next, when a new owner acquires a full portfolio of patents from a party that originally filed a blanket disclosure, it can remain unclear which of these patents is encumbered by licensing obligations. In this case, the ITU/ISO/IEC policy speaks of "reasonable efforts."

In VITA, the general licensing obligation explicitly provides that transferees of patents are to be bound. However, the IPR policy does not provide details on how this obligation is to be satisfied by a member making a transfer. In W3C there are no specific provisions on patent transfers as such. There is, however, a statement that in the case of the acquisition of an entity that is subject to licensing obligations, the obligations will continue to exist. In support of this statement, the document refers to a clause in the W3C policy that specifies that commitments are made "for the life of the patents in question." The policy does not provide details on how this obligation is to be satisfied by a member making a transfer.

The HDMI Forum, which has a recent IPR policy, includes a short clause on patent transfer. When transferring patent claims encumbered by a covenant not to assert (the mode provided for under the HDMI documentation), the owner must ensure that the new owner is also bound to the same licensing commitment. It is not clear whether the commitment will apply across cascading ownership transfers, because the new owner might not be an HDMI Forum member.

For their part, IETF and NFC Forum have no provisions concerning patent transfer at all. An ANSI IPR policy task force is currently considering whether the ANSI patent policy or related guidelines should address the transfer question and, if so, how.

Finally, none of the SSO policies specifies that the organization or other parties need to be notified of ownership changes. Although such changes may become visible to other stakeholders if the new owner submits a licensing declaration, this will not always be the case. In many countries, some but not all ownership changes must be reported to the patent office. Still, patent offices often do not provide mechanisms to make patent ownership changes visible. Moreover, patent transfer recordation practices vary widely across different offices with regard to both what can and must be recorded and how the records may be accessed.

2.6 Summary Observations

Standards-setting organizations have greatly improved the clarity and procedural coverage of their IPR policies from their rudimentary state prior to the late 1990s. In some degree this was forced by legal conflicts, including the

Rambus case, which exposed problems associated with SSOs having unclear IPR policies.[14] Competition among SSOs has also pushed reforms. However, there remain a number of areas of ambiguity and some issues, such as the transfer of licensing obligations when patents are sold, that are only now being addressed. Indeed, there are many other areas that could be analyzed here. However, we focus on three issues.

First, SSOs generally do not define what is meant by FRAND and its component terms: "fair, reasonable, and non-discriminatory." This ambiguity invites increasing litigation in which courts must decide between different parties' widely divergent interpretations of FRAND. Some consortia agree up front on the terms and conditions for a license to essential patents, and it seems that SSO participants could also gain from greater specificity in the context of promoting adoption and implementation of standards. Nevertheless, agreement on a definition of FRAND would be very difficult to achieve in many SSOs for reasons discussed in Chapter 3.

Second, IPR policies still leave a considerable degree of non-transparency regarding how patent information is to be disclosed and used and to whom it is to be made available. Particular elements of policy that could be made more transparent include patent-information updates, the lack of information associated with blanket disclosures, and the failure to make disclosures and licensing commitments public. Publicly available SSO patent disclosure databases—and possibly licensing commitment databases—could help implementers, licensors, and many other stakeholders become more confident about how to secure rights, evaluate claims of essentiality, establish royalty fees if any, and understand competition and antitrust concerns. That said, there are valid reasons why blanket disclosures may be a prudent option, and it is not always clear how and why disclosed information is actually used.

Third, IPR policies of most SSOs do not adequately deal with issues of patent transfers and licensing obligations in today's increasingly dynamic marketplace for patents as assets. The language and mechanisms dealing with this question vary considerably across SSOs. Many policies still appear ambiguous and may not be legally effective in some of the situations they should address, such as those involving licensors and licensees acting in bad faith or in bankruptcy proceedings.

---

[14]*Rambus Inc. v. Infineon Tech. AG.*, 318 F. 3d. 1081, 1097 (Fed. Cir. 2003).

3. Key Issues for SSOs in SEP Licensing

3.1 Introduction

The policies of standard-setting organizations (SSOs) described in the previous chapter address situations in which the SSO develops standards whose use necessarily infringes known intellectual property rights and specify the commitments that participating IPR holders must accept regarding such standards. While the specific language may differ, most SSOs ask rights holders to consent to license their rights on terms that are fair, reasonable and non-discriminatory (FRAND), with or without a royalty payment. For example, the ANSI patent policy requires assurance that either: a) a license will be made available, without compensation, to the applicants desiring to utilize the license for the purpose of implementing the standard; or b) a license will be made available to applicants under reasonable terms and conditions that are demonstrably free of any unfair discrimination. Otherwise, the standard cannot be approved as an American National Standard (ANSI, rev. 2008).

Neither ANSI nor most other SSOs define "reasonable terms and conditions" or the requirements for a license to be "free of any unfair discrimination."[1] Some SSO policies encourage rights holders who have made assurances to reach bilateral agreements with potential licensees. Disputes may be resolved in court or through arbitration or, in some instances, may involve regulatory agencies if there is an allegation of anticompetitive behavior. Despite a relatively small number of cases in which courts have considered the requirements of FRAND licensing terms, there is as yet no broad consensus on this topic.

In this chapter we take a step back and ask the following basic questions:

- What plausible objectives motivate the adoption of FRAND licensing obligations by members of standard setting organizations?
- Do the objectives of members of SSOs differ from societal and competition law concerns?
- How do differences in private and social objectives inform interpretations of FRAND obligations?

---

[1] Special Interest Groups (SIGs) sometimes list explicit licensing terms.

3.2 Objectives of FRAND Licensing Obligations

For virtually all SSOs, the minimum goal regarding IPR is to ensure that all known essential claims in patents are available under FRAND license terms. Some SSOs, or discrete working groups within an SSO, may adopt a more stringent set of rules to seek to have all essential claims made available on a FRAND royalty-free basis. FRAND obligations generally provide assurance that licenses are available for technical solutions involving essential IPR. Participants in SSOs generally do not oppose the development of such standards. For example, the 2008 ANSI Patent Policy states "There is no objection in principle to drafting a proposed American National Standard in terms that include the use of a patented item, if it is considered that technical reasons justify this approach." The FRAND obligations adopted by ANSI and others place limits on the exercise of these patent rights.[2]

As noted earlier, standards-setting organizations have a diverse set of constituents. Some SSO participants are technology owners and users whose business models are based on the sales of products that implement standards and employ patented technologies. Some of these technology users are not interested in asserting standard-essential patents (SEPs) of their own and desire low – or no – royalties for the SEPs they license from others. While these technology users may not proactively seek compensation from implementers for their own FRAND-encumbered patents, they may want their patent rights to have value in terms of offsetting SEPs held by others in the same standards technology area. Other SSO participants are technology sellers whose business models are based on earning royalties from licensing their SEPs to implementers. These enterprises want high royalties and in some cases may want to use their SEPs to demand that implementers accept a license that includes non-SEPs. Still other participants are both technology users and sellers. There are many other business models that impact the views of the various stakeholders.

Technology owners may also adopt different postures depending upon the individual SSOs in which they participate and their particular strategies and goals as they relate to the standard in question. For example, in cases where promoting rapid adoption of a standard enabling a new technology is important to a patent owner and the patent owner can monetize its technology through the sale of products that implement the standard, it may choose to make its SEPs available for free. A patent owner who believes that his SEP is of particular importance to a standard, or one whose technology may meet the needs of a small-volume product, might seek royalties. At a basic level, many SEPs exist for which no license is sought through negotiations unless a third party seeking royalties approaches their owners and implementers. SSOs inevitably shape their

---

[2]We limit the ensuing discussion to patent rights, to which the concept of FRAND obligations uniquely relates. As noted in Chapter 2, IPR policies do typically address copyrights and trademarks as well, but apply a different set of rules that are of limited relevance to this chapter.

IPR policies over time to address the concerns of their existing members and to attract new participants who may be technology users, sellers, or both.

Prevailing practices can also vary from industry to industry, with participants in a number of SSOs in some areas, such as the Internet, preferring to avoid royalty-bearing standards, while those in others, such as consumer electronics, often seek royalties. However, with the convergence of the consumer electronics, information technology, and telecommunications sectors it is becoming more difficult to make distinctions based on the technology area and the impacted industry sectors.

The diversity of actual and potential members of SSOs helps to explain why few of them have developed policies that include detailed definitions of FRAND. Rather, most SSOs rely on general FRAND licensing commitments and certain clarifications with regard to the effect of such commitments as the need arises. SSOs have to govern their IPR policies in an environment of conflicting interests.

To better understand the extent of the limitations imposed by FRAND commitments, it is useful to consider the likely reasons why the members of SSOs arrived at the specific descriptions of their current patent policies. Several concerns are evident.

Ensuring access to patented technologies

Absent a FRAND or other commitment, the owner of a patent has no obligation under the policy to license others to use the patent on any terms. On its face, a FRAND commitment is intended to constrain the freedom that a right holder otherwise has to refuse to license its technology and subsequently enforce its rights. It is understandable that members of a SSO would insist that the organization seek obligations to license patents that are essential to make or use products that comply with a standard. The purpose of an interoperability standard is to coordinate industry activity and take advantage of the economic benefits of scale economies and network effects. These benefits cannot be achieved without widespread licensing of the patented technology that is essential to practice a standard.

Promoting non-discrimination

In the absence of FRAND licensing obligations, there is a risk that implementers will accede to demands to accept discriminatory licensing terms in order to adopt fundamental ICT standards embodying essential patents that cannot be worked around. The owner of a SEP may choose to license the proprietary right with terms that are less favorable to some licensees. For example, without FRAND, a patent owner might license its own customers and partners under terms that differ from those offered to a significant rival, either within or outside the bounds of antitrust law and other legal constraints.[3] Thus, an objective of a FRAND licensing commitment is to avoid discriminatory licensing terms that disadvantage some licensees by imposing on them substantially larger royalties or more restrictive conditions in comparison with others. Indeed, this assurance of non-discrimination is fundamental to the weighing of benefits and costs that an industry participant makes when deciding whether to participate in an SSO or adopt a standard.

Avoiding ex post hold-up

The economic alternatives available to licensees often differ before and after the adoption of a standard. In the process of developing a standard, several alternative technological solutions may be available that have similar cost and performance characteristics. "*Ex post*," after firms and consumers make investments that are specific to the standard, the economic choices are far more limited if adopting an alternative technology for the standard would impose substantial additional costs and delays. Furthermore, in markets with large network externalities it may not be feasible to coordinate the actions necessary to switch the market to a different standard. *Ex post* hold-up can occur if the owner of a SEP chooses royalty terms that reflect the high cost of switching to an alternative technology after firms and consumers have made specific investments, rather than the value of the claimed invention.

Ex post patent hold-up imposes costs on licensees and consumers. It also rewards the patent holder via windfall profits reflecting the costs of switching to an alternative technology rather than the economic merit of the selected standard. Because such hold-up is potentially costly to members of SSOs, it is reasonable for their IPR policies to seek to guard against it by requiring FRAND commitments. This concern is greatest in ICT sectors where standards aspire to global adoption, switching to alternative standards can be costly, and patent filings have proliferated in many countries.

Competition authorities in both the EU and the United States see the potential for costly hold-up, as noted in the following joint statement by officials of the Department of Justice Antitrust Division, Federal Trade Commission, and Competition Directorate General of the European Commission:

> SSOs constrain the license terms for SEPs because of the substantial market power necessarily enjoyed by the owner of an SEP in a successful standard. Moreover, this market power is achieved through the joint action of entities—the SSO members—that might be in competition with each other outside the SSO (Kühn, et al., 2013).

---

[3] *Georgia-Pacific Corp v. United States Plywood Corp.,* 318 F. Supp.1116 (S.D.N.Y. 1970). The well-known *Georgia-Pacific* case, in fact, recognizes the relationship between the patent owner and the infringer as a factor to consider in assessing "reasonableness" in a patent damages case.

Managing royalty allocation and stacking

Interoperability standards often have numerous patents declared essential to their use. Economic valuation of a particular patent becomes difficult and even arbitrary in this circumstance. If several patents are essential to make or use a product that complies with a standard, then each patent has a claim on the value of the product. There is a concern that, acting independently, the individual holders of patents essential to a standard will demand royalties that, in the aggregate, are so high as to impede the adoption and use of products that implement the standard. Such high aggregate royalties can be detrimental to rights holders as well as to the consumers of products that implement the standard and can be a drag on future innovation. Economists use the term "royalty stacking" (Lemley and Shapiro, 2007) to describe outcomes in which the cumulative effects of individually rational royalty demands result in aggregate royalties that harm consumers, rights holders and innovators.

Royalty stacking is different from standard-related hold-up. When a standard-using product requires numerous SEPs, a single patentee can demand a disproportionate share of product value even if licensees do not incur costs to switch to a different product. Because all the SEPs are necessary, licensees and customers would switch to a different product only if the total royalty is excessive, which leaves room for individual patent holders to demand a disproportionate share of total royalties.

Several important ICT standards illustrate the potential for royalty stacking. GSM is a second-generation standard for mobile telecommunications and W-CDMA (also called UMTS) is a third-generation standard for networks based on the GSM standard. One study identified more than 50 entities that each disclosed patents or patent applications as essential to the GSM or W-CDMA standards.[4] Together these entities disclosed over 23,500 patents, belonging to at least 1729 patent families.[5] Other mobile telephony technologies have similar characteristics. Numerous firms in total declared more than 750 unique patent families as essential to the second-generation GSM standard and 500 unique patent families as essential to the fourth-generation LTE standard.[6]

A U.S. district court concluded that 92 companies identified patents as essential to the 802.11 ("Wifi") wireless local access network family of standards and 59 companies filed blanket declarations without identifying specific patents. The court accepted testimony that there are possibly thousands of patents declared essential to the 802.11 family of standards.[7] The same court concluded that approximately 33 U.S. companies declared patents essential to the H.264

---

[4] See Bekkers and Martinelli (2012), Table 6, p. 1205. (The table includes GSM as well as W-CDMA patent disclosures).

[5] Bekkers and Martinelli (2012), pp. 1203-1205.

[6] See Blind et al. (2011), Table 3-3, p. 36.

[7] *Microsoft Corp. v. Motorola, Inc.*, Findings of Fact and Conclusions of Law, 2013 U.S. Dist. LEXIS 60233 at paragraph 335 (W.D. Wash., Apr. 25, 2013).

advanced video coding standard and 19 additional companies provided blanket declarations to the ITU (one of the developers of the standard) without identifying specific patents.[8]

These figures could both overstate and understate the potential for royalty stacking in different respects. The numbers could overstate the potential for royalty stacking because patent owners that participate in SSOs have incentives to declare, as essential, patents that may not in fact be valid or infringed by a product that implements the standard. Evaluating the scope and validity of every patent is a costly exercise and patent holders may prefer to declare patents as essential without subjecting them to careful scrutiny. Further, SEP owners may err on the side of disclosing patents that are not essential to the standard because failure to disclose a SEP for which royalties are demanded, where it is required by an SSO, may expose the owner to future litigation. Finally, ownership of a large stock of patents that are declared essential to a standard can be a valuable asset for a firm that seeks licensing royalties or cross-licenses at favorable terms. A larger quantity of essential patents is also more valuable if the patents are contributed to a patent pool that seeks royalty income and allocates that income in proportion to the number of patents in the pool.

Estimates of essential patents and the number of SEP owners could understate the potential for royalty stacking in other respects. Multiple SSOs participate in the development of some standards and not all SSOs maintain easily accessible databases of patent declarations. Some participants in SSOs make blanket disclosures and commit to license patents deemed essential to a standard at FRAND terms without identifying the specific patents in advance. Also, some firms might hold essential IPR without having an obligation to disclose these patents, for instance because they are not participating in the Working Group in question or because the participating individuals were not aware of those patents. Non-participant firms are not required to disclose the existence of these patents and would not appear in a review of licensing commitments at the relevant standard-setting organizations.

Regardless of whether observed statistics overstate or understate the number of owners of essential patents, it is certain that numerous distinct entities own patents that are essential to make or use products that implement many common ICT standards. Given the large number of SEP owners, efforts to obtain royalty income for these patents can result in heavy monetary burdens for those who make or use products that implement the standard. In theory, the total royalty stack can be so large that it would suppress the adoption or use of standardized technologies and impose excessive costs on all segments of the industry that implement a standard.

Despite this concern, the committee has found no empirical evidence showing that royalty stacking currently suppresses the adoption or use of standard-compliant products. Firms can mitigate the burden from aggregate royalties

---

[8]*Ibid.*

in several ways.[9] Some rights holders enter into cross-licensing arrangements with zero royalties or with royalties equal to the difference between the fees charged by each party. Patent pools exist for some standards, which reduce transaction costs and mitigate royalty stacking by setting a single fee for a portfolio license. Firms that own patents and sell products covered by those patents have incentives to charge low or zero royalties to promote the commercialization of their products. In addition, firms have strategic incentives to refrain from charging high royalties. Indeed, product prices have been dropping for devices such as mobile phones and laptop computers that support multiple standards for which there are thousands of declared SEPs owned by hundreds of entities. Furthermore, not all standards, even in the ICT area, invoke large numbers of patents with widely distributed ownership.

Nonetheless, the committee cautions that the costs from royalty stacking could increase in the future if more patent owners choose to monetize their patent rights. At some point the cumulative burden of making multiple royalty payments to distinct entities could become so large that adoption or utilization of standard-compliant products would be suppressed and the resulting higher costs of developing and producing these products may become a drag on future innovative efforts.

Courts play a crucial role in this area since judicial decisions both directly set norms for FRAND terms and settle disputes over whether royalty offers comply with FRAND commitments. Thus, courts can contain the risk of royalty stacking and hold-up by ensuring that awards for patent infringement are reasonable, taking into account the contribution of the patented inventions and the costs of obtaining all other intellectual and physical inputs that are necessary to make or sell the infringing product.[10] For patents that are essential to a standard and allegedly infringed by an implementing product, this requires allocation of the value contributed by the standard as opposed to the contributions of others, including patents that are essential to other standards.

An allocation of the value of a standard to its essential patents based on simple numeric proportionality, under which a patent owner's share of value is equal to its share of patents that are essential to make or use the products at issue, has the virtue of simplicity and ease of administration. However, it also raises the potential for imprecision and strategic manipulation. Some patents are more valuable than others in terms of their contributions to the standard, either because their validity has been firmly established through litigation or because

---

[9]See Damien Geradin, Anne Layne-Farrar, and A. Jorge Padilla, The Complements Problem Within Standard Setting: Assessing The Evidence On Royalty Stacking, *B.U. J. Sci. & Tech. L.*, Vol. 14:144-176.

[10]Additional claims on the value of the product could come from patents covering other standards implemented in the product and patents on proprietary technologies that contribute to the overall value of the product.

they are more central to implementing products.[11] Furthermore, a numeric proportionality rule encourages patent owners to file separate patents for essential claims instead of including multiple claims in a single patent. Nevertheless, the fact that many owners of SEPs participate in patent pools that allocate royalties based on numeric proportionality suggests that such rules, however imperfect, are a plausible starting point for the valuation of individual SEPs.

On April 25, 2013, Judge James L. Robart in the Western District of Washington issued a detailed decision calculating FRAND royalty rates for products that involve multiple standard-essential patent rights.[12] The decision relied on the factors laid out in the earlier *Georgia-Pacific* case[13] but modified them to account for the FRAND context. In particular, Judge Robart noted that in considering a hypothetical negotiation to arrive at a reasonable royalty "the hypothetical negotiation almost certainly will not take place in a vacuum: the implementer of a standard will understand that it must take a license from many SEP owners, not just one, before it will be in compliance with its licensing obligations and able to fully implement the standard."[14] In particular, Judge Robart concluded that "a proper methodology for determining a [F]RAND royalty should address the risk of royalty stacking by considering the aggregate royalties that would apply if other SEP holders made royalty demands of the implementer."[15]

At issue in the case was the value of Motorola's patents for Microsoft products that implemented the ITU H.264 standard for video processing and the IEEE 802.11 family of standards for wireless communications. In constructing a reasonable royalty, Judge Robart focused on royalties that Motorola would have earned for its patents if they had been contributed to existing patent pools relating to those standards. Although the judge listed a number of reasons why pool rates should not be determinative of the FRAND rates for all SEPs for a standard, he found them to be useful indicators based on the facts of the case.

Judge Robart's approach provides less quantitative utility to assess reasonable royalties for multiple standard-essential patents where there are no existing patent pools of SEPs related to the standard in question. Nonetheless, his efforts to distinguish comparable royalty negotiations, identify the risks of royalty stacking, and take into account the potential aggregate burden of licensing

---

[11] For illustrations of royalty allocations under different assumptions, see Layne-Farrar et al. (2007) and Salant (2007). Although several techniques are available to allocate value to SEPs or to patents that are complementary to the value of products that implement a standard, they require more information about technology characteristics than is typically available.

[12] *Microsoft Corp. v. Motorola, Inc.*, Findings of Fact and Conclusions of Law, 2013 U.S. Dist. LEXIS 60233 (W.D. Wash., Apr. 25, 2013) [hereinafter *Microsoft v. Motorola FFCL*].

[13] *Georgia-Pacific Corp. v. United States Plywood Corp.* (318 F. Supp. 1116 [S.D.N.Y. 1970]).

[14] *Microsoft Corp. v. Motorola* FFCL, op. cit. at 11.

[15] *Ibid.*

demands from many patent owners provide guidance that may be useful to assess lower and upper bounds for FRAND royalties when many patents are essential to make or use products that comply with a standard.

In general, determining whether a given licensor is making excessive demands that contribute to either royalty stacking or standard-related hold-up can be very challenging. If a standard requires the patented contributions of many different licensors, each of which is essential to implement the standard none of which have close substitutes *ex ante* or *ex post*, then the division of royalties among patented contributions is essentially arbitrary. In other instances of standard setting, some features of a standard may be more important than others and some features (and the technologies that implement them) may have feasible substitutes prior to the standard being adopted. In these substitutions some licensors may argue, with some justification, that their technology is worth more than others. Further, in the presence of switching costs it can be difficult or even impossible to distinguish excessive demands from hold-up. A large royalty demand could reflect the effort of a patent owner to appropriate a share of the switching costs that lead to *ex post* hold-up. But it also could be an effort by a particular patent owner to capture a disproportionate share of an *ex ante* reasonable aggregate royalty.

There is little evidence that the existing IPR polices of most large SSOs effectively limit the ability of individual patent owners to negotiate for a disproportionate share of product value in their royalty demands. Still, some SSOs, including various consortia, IEEE, Wi-Fi, and ETSI, have explored the idea of using various licensing disclosure arrangements to attempt to avoid royalty stacking, although the effort and expense of putting such arrangements together is justified only in some circumstances.

One approach that may address concerns about royalty stacking is for SSOs to require patent owners that intend to assert their patents to post a maximum royalty before the standard is adopted. The posting of royalties would allow potential licensees to detect potential *ex ante* hold-up and possible royalty stacking by considering the implications of individual posted maximum royalties for the aggregate royalties required to make or use products that comply with the standard. SSOs might also establish mechanisms for effectively avoiding or resolving disputes, such as alternative dispute resolution (ADR).

That said, policies requiring *ex ante* disclosure of licensing terms have been suggested and rejected at several SSOs. Such policies are unpopular with patent holders that prefer to negotiate royalty rates with potential licensees and not to be pressed into royalty rates and terms before the standard is finalized and relevant SEPs can be determined.

Another objection to *ex ante* licensing disclosure policies is that they could lead to anticompetitive conduct by licensees if royalty rates were jointly negotiated—a scenario referred to as oligopsony or a buyers' cartel. Given this concern, both VITA and IEEE obtained written assurance from competition authorities before adopting *ex ante* licensing disclosure policies.[16] Opponents of *ex ante* disclosure of licensing terms also point to practical difficulties, such as fixing the timing of disclosures and designing policies that would accommodate blanket disclosures and avoid the need for costly patent searches.[17] The latter concern is especially relevant for companies that generally do not proactively seek licenses from implementers, but rather use their SEPs largely for defensive purposes. In addition, many SEPs arise not because a patent holder contributed the technology to the SSO, but rather as a result of the collective drafting exercise. Mandatory *ex ante* disclosure could also disrupt technical committee work, if participants were asked to review all possible SEPs and related licensing terms, particularly for ICT standards that can reach hundreds of pages in length and implicate dozens or hundreds of patents.

Because few SSOs have adopted policies with regard to *ex ante* disclosure of licensing terms, empirical evidence on these questions remains quite limited. However, one study found that the *ex ante* disclosure policies adopted by VITA, where disclosure is mandatory, and IEEE, where disclosure is optional and seldom used, had no measureable impact on their standard-setting processes (Contreras, 2013).

Another proposal to address royalty stacking involves an SSO's establishment of an aggregate cap on royalties that can be charged with respect to patents essential to a particular standard. This approach was proposed within ETSI as early as 2005, but its consideration was terminated after an unfavorable reaction by the European Commission's Competition Directorate-General.[18] Yet another possible approach would be for SSOs or courts to make an effort to consider the landscape of patents that are essential for use of a standard. That recognition would help avoid some of the most egregious errors (such as 5% royalty per patent). Equal apportionment of values, though sometimes flawed, could be a starting point from which to argue that some patents are worth more than others, while still recognizing that one or a few patents may not account for all or most of the product value when there are many other essential patents.

---

[16] See U.S. Department of Justice, Business Review Letter to VMEbus International Trade Association (Oct. 20, 2006), U.S. Department of Justice, Business Review Letter to Institute of Electrical and Electronics Engineers (Apr. 30, 2007), and Deborah Platt Majoras, Recognizing the Procompetitive Potential of Royalty Discussions in Standard-Setting Remarks prepared for "Standardization and the Law: Developing the Golden Mean for Global Trade," Stanford Law School (Sept. 23, 2005).

[17] These difficulties apply generally to SSO policies regarding *ex ante* disclosure of SEPs, not just to licensing terms, as discussed in Section 4.

[18] Claudia Tapia, Industrial Property Rights, Technical Standards and Licensing Practices (Frand) in the Telecommunications Industry 165-66 (2010). A modified form of this proposal has recently been made by Contreras, who analogizes the aggregate royalty on SEPs covering a particular standard to the collective royalty charged by a patent pool. See Jorge L. Contreras, Fixing FRAND: A Pseudo-Pool Approach to Standards-Based Patent Licensing, 79 Antitrust L.J. (2013).

3.3 Interpretation of FRAND Obligations to Address Competition and Efficiency Concerns

Competition agencies in the United States and the European Union have proposed interpretations of FRAND licensing obligations in an attempt to "fill in the gap" created by the lack of clarity in SSO IPR policies.

Incremental value

Competition authorities and a number of scholars have endorsed the principle that a "fair and reasonable" royalty should reflect its incremental value relative to the next-best alternative assessed before firms and consumers make investments specific to the technology. For example, the Federal Trade Commission concludes that "Courts should cap the royalty at the incremental value of the patented technology over alternatives available at the time the standard was chosen." (U.S. Federal Trade Commission, 2011) Guidelines issued by the European Commission state that "whether fees imposed for patents in the standard-setting context are unfair or unreasonable will be based on whether the fees bear a reasonable relationship to the economic value of the patents." In making this assessment, the Commission notes that "it may be possible to compare the licensing fees charged by the undertaking in question for the relevant patents in a competitive environment before the industry has been locked into the standard (*ex ante*) with those charged after the industry has been locked in (*ex post*)." (European Commission, 2010)

Many SSOs emphasize that it is desirable to have early disclosure of relevant patents and some have provided for *ex ante* commitments to specific licensing terms and conditions. However, absent further clarification of the meaning of FRAND, it is not clear whether members of SSOs intend that FRAND royalty commitments should reflect incremental values or some other notion of fair and reasonable pricing. This topic is currently under discussion at some prominent SSOs. If the term does not reflect incremental value one might question whether norms such as economic efficiency should determine the interpretation of fair and reasonable license terms.

Incremental value provides a means to assess the *ex ante* contribution of a patent that covers a discrete technology whose value can be assessed independently from the contributions of other technologies. For example, a patent on a technology to increase the signal-to-noise ratio of transmissions using a wireless communications standard can be assessed independently from the contributions of other patents that are essential to the standard. Farrell et al. (2007) provide a formula for the incremental value of a technology. Swanson and Baumol (2005) suggest that SSOs should conduct an auction to determine the incremental value of the best technology. These approaches may provide feasible means to estimate the value of a patent that is essential to a standard that involves a single patented technology and if alternative technologies can be compared based on a single attribute such as cost. However, the technologies incremental approach is not sufficient to determine the appropriate royalty for a particular patent when there are many SEPs, as is typically the case for interoperability standards.

When multiple patents are essential to make or use products that comply with a standard, neither incremental value nor auction approaches provide a practical means to allocate the economic value of the technology to different necessary patents within the standard. When patented technologies in a standard have different attributes, valuation requires aggregation of the attributes into a metric that can be used to rank alternatives. This is a complex problem and technology users are likely to disagree about the appropriate weights for the different attributes.[19]

Practically speaking, in assessing "reasonableness," courts have resorted to measures other than incremental value. Recently, a long-time rule of thumb gauging royalties at 25% of the infringer's profits was rejected by the U.S. Court of Appeals for the Federal Circuit.[20] In accordance with a statement by the European Commission, the royalties a patent has realized in the past before a standard is approved can shed light on value.[21] What royalties the patent has realized in the past in other license negotiations after or (in accordance with the European Commission statement) before a standard is approved, can shed light on value. Royalties for comparable patents or licenses that are encumbered by FRAND commitments may also inform the evaluation of reasonable royalties, although recent cases have raised the bar on validating the circumstances and expert testimony on which cases are comparable.[22]

At the most basic level, a commitment to license at RAND terms should not permit a patent holder to obtain a royalty that reflects standardization effects rather

---

[19] Patent pools have developed "rough and ready" approaches to this apportionment problem, such as allocating royalties in proportion to each firm's count of essential patents. However, each approach bears its own problems, especially where attributes must be weighted.

[20] *Uniloc USA Inc. v. Microsoft Corp.* 10 1035 (U.S. Court of Appeals for the Federal Circuit, 2011).

[21] See *Communication from the Commission: Guidelines on the applicability of Article 101 of the Treaty on the Functioning of the European Union to horizontal cooperation agreements*, at http://eur-lex.europa.eu/LexUriServ/LexUriServ.do?uri=OJ:C:2011:011:0001:0072:EN:PDF. As an economic matter, it is questionable whether ex post royalties should inform FRAND royalties.

[22] As noted by Judge Robart in *Microsoft Corp. v. Motorola, Inc.*, No. C10-1823JLR (U.S. District Court, 2013), not all royalties are comparable or based on similar considerations. For example, "[t]he court concludes that where multiple technologies (including both standard essential and non-essential patents) are licensed within the same agreement, it is necessary to apportion the value of [the SEPs at issue] from the other licensed properties. Such apportionment would be difficult." (Order 137-138). See also *Uniloc USA, Inc. v. Microsoft Corp.*, 632 F. 3d 1292, 1317 (Fed. Cir. 2011) ("The meaning of these cases is clear: there must be a basis in fact to associate the royalty rates used in prior licenses to the particular hypothetical negotiation at issue in the case.").

than the value of the claimed invention; nor should a patent holder automatically realize less than the rate based on the *ex ante* contribution of the invention merely because s/he participates in a standards effort. In *Microsoft v. Motorola*, Judge Robart articulates two high-level principles for determining FRAND rates. First, the value of the SEPs should be assessed separately from their inclusion in the standard. Second, a FRAND rate should be based on the importance of the patents to the standard and the importance of the standard and the patents to the products at issue. As noted above, one corollary of this second principle, the ruling suggests, is that when there are multiple holders of FRAND-encumbered SEPs for the same standard, it would be necessary to consider cumulative royalty rates.

U.S. courts frequently cite the 15 factors enumerated in *Georgia-Pacific Corp. v. United States Plywood Corp.*[23] that are relevant to a hypothetical negotiation between a patentee and a licensee to assess monetary compensation for patent infringement. A key issue now is whether these factors are informative for assessing FRAND royalties for standard-essential patents. Judge Robart concluded that they are informative, albeit with important qualifications. He noted that the hypothetical negotiation under a FRAND obligation differs from the typical *Georgia-Pacific* analysis conducted by courts in a patent infringement action because, among other reasons, the owner of a SEP is under the obligation to license its patents on FRAND terms.

For example, the first *Georgia-Pacific* factor is "The royalties received by the patentee for the licensing of the patent in suit, proving or tending to prove an established royalty" and the second factor is "The rates paid by the licensee for the use of other patents comparable to the patent in suit." Judge Robart concluded that to be comparable, past royalty rates for a litigated SEP or another similar patent must be negotiated under a clearly understood FRAND obligation.

Several *Georgia-Pacific* factors address the technical characteristics of the patented technology and its value to the licensee. Factors 6 and 8 relate to the value of the patent in promoting sales of the licensee's products and factor 9 is "The utility and advantages of the patent property over the old modes or devices, if any, that had been used for working out similar results." Factor 10 is "The portion of the realizable profit that should be credited to the invention as distinguished from non-patented elements, the manufacturing process, business risks, or significant features or improvements added by the infringer." In interpreting the implications of these factors for a hypothetical FRAND royalty negotiation, Judge Robart concluded that a reasonable royalty for a SEP should reflect the contribution of the patent to the standard and to the value of the implementer's products, but should not take into account the value to the licensee created by the standard itself.

Factor 15 is "The amount that a licensor, such as the patentee, and a licensee (such as the infringer), would have agreed upon at the time the infringement began if both had been reasonably and voluntarily trying to reach an agreement ..." In evaluating the implications of this summary factor, Judge Robart noted

---

[23]318 FSupp 1116, 6 USPQ 235 (SD NY 1970).

that "the SEP owner would have been obligated to license its SEPs on RAND terms which necessarily must abide by the purpose of the RAND commitment of widespread adoption of the standard through avoidance of holdup and stacking."

It remains to be seen whether other courts will adopt the FRAND framework described by Judge Robart in *Microsoft v. Motorola*. Nonetheless, the court's application of the *Georgia-Pacific* factors to a hypothetical negotiation over royalties for a FRAND-encumbered patent emphasizes certain points that are widely discussed in the literature and have achieved some consensus within the standards community and among antitrust authorities. Specifically, the court determined that the participants in a "hypothetical negotiation would set [F]RAND royalty rates by looking at the importance of the SEPs to the standard and the importance of the standard and the SEPs to the products at issue."[24] It is reasonable to apply the *Georgia-Pacific* factors to help determine FRAND licensing rates, with appropriate modifications to reflect the commitments made by SEP owners and the characteristics of industries within which SEP owners and implementers operate.

Non-discrimination

Some notion of non-discrimination must be central to any meaningful FRAND concept.

In his review of the *Georgia-Pacific* factors to assess FRAND royalties, Judge Robart noted that the fifth factor—"The commercial relationship between the licensor and licensee, such as, whether they are competitors in the same territory in the same line of business; or whether they are inventor and promoter"—does not apply in the FRAND context. This is because having committed to license on FRAND terms, the patentee is obligated to license all implementers on reasonable terms and may not discriminate against its competitors in terms of licensing agreements.

Although the non-discrimination requirement of the FRAND commitment is clear, it is much less clear what that implies in practice. A patentee can structure royalties in different ways. One might imagine a licensor asking for a unitary fixed charge independent of the licensee's sales volume, a given fee per unit of sales, or a mixed royalty. Each of these may be considered non-discriminatory on its face but generate wide variations in effective licensing terms across licensees. For example, a royalty rate that is a fixed percentage of a using product's market value implies a much higher per-unit dollar royalty on higher value goods than on less expensive goods, even though the same technology is in play.

In general, neither a single fixed fee nor a single per-unit fee is likely to promote the most efficient utilization of a technology. Economic output may increase if licensees are free to choose among different fee schedules. For example, a licensee may choose a royalty schedule that has a low initial fee and a

---

[24]*Ibid.*, p. 4.

high per-unit fee or a different royalty schedule that has a higher initial fee and a lower per-unit-fee.

One interpretation of the non-discrimination requirement is that patent owners should offer licensees the same choices of royalty schedules. Under this interpretation, the alternatives of a high initial fee/low per-unit fee or a low initial fee/high per-unit fee generally would not be discriminatory if these choices are available to all licensees. It is not clear, however, whether this interpretation is consistent with the non-discrimination requirement of FRAND or whether non-discrimination requires such menus. Moreover, there are circumstances in which such choices likely should be considered discriminatory. For example, the choices could be structured so that only very large users of a technology could profitably accept a license.

Portfolio licensing and cross-licensing

The ability to determine whether the price and other terms relating to a given SEP or essential claim are reasonable and have been applied on a consistent basis is further complicated by the fact that those claims may be part of a bundle of other claims. The latter may not be essential to the implementation of the standard but could be needed in connection with the manufacture of any product incorporating the feature that complies with the standard. Typically, if both parties agree, all of these claims will be included under a single license that includes a single set of terms applicable to all of the referenced patents or claims. Because the owner of these claims is under no obligation to license the non-essential, but desirable, claims included in the bundle subject to FRAND terms, it will be difficult to determine whether a FRAND obligation has in fact been met.

For many companies the objective of taking a license is to obtain the freedom to operate in a technology space. These companies may not be interested in a license to a single patent if their activities may infringe other patents owned by the licensor. Instead, they may want access to the portfolio of patents owned by the licensor, including future patents. This can be achieved with an appropriately structured portfolio license. Similarly, the licensor often may want the freedom to operate without the risk of liability for infringing the licensees' current or future patents. This freedom can be obtained with an appropriately structured cross-license between the two companies.

Portfolio licenses and cross-licenses raise challenges for evaluating whether a particular patent license is consistent with a FRAND obligation. The royalty paid for a patent portfolio covers many patent licenses, and it can be difficult, or even impossible, to allocate the portfolio royalty to individual patents in a meaningful way. Cross-licenses add a further complication because parties to the arrangements often net out the payments for patents they entail. Parties to a cross-license may not pay any royalties if they agree that their respective portfolios have equal values. But that does not mean that the portfolios—or the individual patents in the portfolios—have no values.

Portfolio licenses and cross-licenses raise issues of transparency for patent royalties. It is difficult to know what the royalty may be for a single patent in a portfolio license or a cross-license. If the SEP owner insists that the SEP be licensed as a component of a portfolio that includes other, non-essential patents, the result can be a demand for a royalty in excess of the FRAND rate for the SEP. A similar outcome may ensue if the licensor insists on a cross-license to patents that have a total value in excess of the FRAND royalty on the SEP. Of course, licensees may not agree to such demands in practice.

Mutually agreed cross-license arrangements are certainly acceptable under SSO policies. However, one-sided demands by SEP owners that the licensee accept patents other than the SEPs in the standard, as a condition of access to the SEPs, would violate the terms of many IPR policies. These expressly limit cross-license terms, typically under a "reciprocity" provision, to other claims that are essential to the implementation of the same standard. An owner of SEPs in this situation would not be permitted to demand a broader cross license without violating its FRAND obligation. Even absent such an explicit rule, trade usage might indicate that the party demanding the broad cross-license would be found in court to have violated its FRAND commitment. The competition authorities have suggested that it likely would be anti-competitive for a holder of FRAND-encumbered SEPs to demand a cross-license to non-SEPs unless both parties voluntarily agree to a broader cross-license deal.

The Department of Justice has suggested that SSOs should prohibit the mandatory cross-licensing of patents that are not essential to the standard or a related family of standards, while permitting voluntary cross-licensing of all patents. Further, in its Decision Regarding Google's Acquisition of Motorola Mobility DG Competition of the European Commission wrote that "Another concern would be that the SEP holder may force a holder of non-SEPs to cross-license those non-SEPs to it in return for a license of the SEPs."[25]

The problems of low transparency and potential discrimination in licensing terms could be addressed by requiring owners of patents with FRAND obligations to provide to an implementer, upon request, a royalty or royalty schedule at which they are willing to license their patents (Gilbert, 2011).[26] Portfolio licensing and cross-licensing would remain permissible choices. However, parties to a portfolio or cross-license would have the alternative of accepting a license for a single patent at the posted rate.[27]

Other licensing terms

Licenses typically include many terms negotiated between parties. In addition to fixed and per unit royalties, the license may have a royalty cap, payments

---

[25]See European Commission (2012).
[26]The committee cautions that there is little systematic evidence of how such a requirement would affect incentives of inventors to contribute IP to a developing standard.
[27]See for example, American Bar Association (2007).

that are conditioned on the licensees' sales, and discounts for prompt payment. The royalties comprise only one component of licensing terms. The license may be worldwide or restricted to a country or region. The license may be for a particular field of use or for any economic activity by the licensee. Royalty terms may differ depending on the products made using the licensed technology.

In addition, license agreements may seek to include grant-back provisions in which the licensee agrees to provide the licensor with a license to certain future patents issued to the licensee. In fact, most SSOs' IPR policies are silent on the topic of grant-backs, which may or may not be within the FRAND framework. Where it exists, the relevant IPR policy may limit the scope of the grant-back to patents that are essential to the same standard and may address whether the grant-back is exclusive or non-exclusive and whether it has a royalty obligation.[28] The grant-back license may require that the licensee agree either to license its SEPs under FRAND terms or not to assert its own SEPs under the same standard against the licensor or against other licensees implementing the same standard.

This complexity of licensing terms greatly complicates the determination of "fair and reasonable" royalty terms and any assessment of unfair discrimination.[29] A license with a given per-unit royalty that is limited to production in Europe and requires a royalty-free grant-back of future patents is not comparable to an unrestricted worldwide license with the same per-unit royalty. While the IPR policies of SSOs could theoretically define a standard set of licensing terms to facilitate compliance with FRAND licensing obligations, this option has rarely been pursued. One reason is that SSO members prefer to sell technology under their own licenses.

The most common exception arises where the founders of the standardization effort seek to develop only one or a few closely related standards. In this case, the founders will often enter into a cross-license among each other, commonly referred to as a "Promoters' Agreement," under which the SEPs of all promoters are pooled on a FRAND or FRAND-RF basis. Under some Promoters' Agreements, each promoter in turn may be free to license the pool of claims to third parties for purposes of implementing the standard in question. These rights are contained in "Adopter Agreements," which may include royalty-free grant-back terms similar to those in place between the Promoters. Under such an arrangement, extraneous licensing terms are effectively excluded.

As Judge Robart concluded, FRAND obligations may be informed when a formal patent pool is formed, under which implementers pay a single fee referenced in a common fee schedule and sign virtually the same license as all other pool licensees. Because such pools are difficult and expensive to create and of

---

[28]Grant-backs are generally non-exclusive since competition questions could arise where the licensor precludes its licensee from transacting with other firms.

[29]This statement should be qualified by noting that the patent holder's commitment to license on fair and reasonable terms does not prohibit it from entering voluntarily into more complex but mutually beneficial licensing arrangements.

limited value if they do not attract very wide participation either by SEP holders or interested implementers, they are rarely created. But they can serve a valuable purpose in setting FRAND rates, providing guidance in structuring FRAND licenses, and potentially addressing patent-stacking issues.[30]

Summary

There are many respects in which the private concerns of members of standards-setting organizations are congruent with the objective of creating standards that enhance economic welfare. SSO members want to develop standards that are effective technical solutions. They want to avoid capture of a standard by one or a few entities, and they want a fair opportunity to participate in setting the standard. These are all consistent with enhancing economic welfare.

However, SSO members do not necessarily internalize all of the relevant effects of their standards choices and IPR policies on economic welfare. SSO members who are implementers may not decisively oppose high royalty payments, even for patents of questionable technical value in the standard, if they can pass these costs onto consumers. Moreover, they may face the threat of the SEP holder obtaining injunctive relief if they do not accede to the latter's licensing demands.[31] High royalties can reduce output and raise product prices, which harm consumers. To be sure, these costs should be set against the dynamic benefits of royalty income. Patents are an important means of realizing returns on R&D investment. Royalties are often used to fund further research that develops better and more efficient products and improved methods of achieving interoperability.

Despite the lack of clear guidance to date in SSO IPR policies, we should not understate the importance of a FRAND commitment and its ability to deal with potential strategic concerns. SSOs are largely effective by virtue of the credibility of the voluntary consensus-based standards process in delivering high quality standards that can be implemented while avoiding serious IPR-related consequences. SSOs remain important and the fact that more are created all the time suggests that market participants know they have much to gain by expending resources in helping develop and implement standards. Indeed, alleged abuses of FRAND commitments have been relatively infrequent in the past relative to the large number of standards and the much larger number of SEPs that standards incorporate. This suggests that most rights holders have not pushed the limits of a FRAND interpretation through the strategic enforcement of their patents.

Within the past few years, however, there have been an increasing number of lawsuits alleging that SEP holders have demanded non-FRAND terms from implementers and used the threat of injunctive relief to try to force these implementers to accept such terms or risk having their product sales halted. This may

---

[30] These points were addressed also in Judge Robart's opinion, cited above at footnote 37.

[31] See the related discussion in Chapter 6 on injunctions.

Key Issues for SSOs in SEP Licensing 69

signal that attempts at patent hold-up are increasing, or it may be that in the past implementers have chosen to accede to unreasonable terms instead of challenging them in courts.[32] The current debates at a number of SSOs about whether to further clarify the effect of a FRAND licensing commitment in their IPR policies suggest that industry players have diverse views as to what FRAND means and what constraints it actually places on the SEP holder.

Because of the strategic concerns listed earlier, competition authorities and a number of companies now recommend that SSOs clarify the various effects of a FRAND commitment by formulating certain principles. These principles may include, among other conditions for compliance with FRAND, guidance related to royalty demands that could be a disproportionate share of product value when many patents are necessary to comply with a standard.

3.4 Recommendations to SSOs

The issues discussed in this chapter are highly complex, leaving relatively little room for broadly applicable recommendations. However, there are areas in which the committee sees the potential for improvement relating to SSO policy regarding licensing commitments. Underlying these recommendations is a fundamental principle: The committee believes that a FRAND licensing commitment represents more than the patent owner offering a license on its own terms. A FRAND commitment is also mutual in the sense that both the SEP holder and any prospective licensee are expected to negotiate in good faith towards a license on reasonable terms and conditions that reflect the economic value of the patented technology.

Recommendation 3:1

The committee urges SSOs to become more explicit in their IPR policies regarding their understanding of and expectations about FRAND licensing commitments. SSOs should clarify the various effects of a FRAND commitment by formulating certain statements of principle. These principles could include, among other conditions for compliance with FRAND, guidance regarding royalty demands that could be a disproportionate share of product value when many patents are necessary to comply with a standard and the relevant product includes multiple technologies.

Recommendation 3:2

The committee recommends that SSOs include in their policies statements that implementers and the consumers of their products and services are the in-

---

[32]A further factor in mobile communications may be that a number of large participants contribute relatively little to standards development and may not want to pay for the cost of the innovation underlying those standards.

tended third party beneficiaries of licensing commitments made by SSO participants. Although the enforceability in all courts of such a term may not be guaranteed as the law in this regard is still evolving, inclusion of such statements would inform courts of the intent of SEP owners participating in SSO working groups. It would also provide greater confidence to potential implementers and promote greater certainty in the event of a dispute.

Several recommendations are aimed at improving clarity within SSOs regarding the bundling of licensing commitments.

Recommendation 3:3

SSOs should clarify in their policies that prospective licensees may request a license to some or all FRAND-encumbered SEPs owned or controlled by a patent holder. Licensors may not tie the FRAND commitment and the availability of the requested SEPs to a demand that a licensee accept a package or portfolio license that includes non-SEPs or SEPs for unrelated standards. Nor may the licensors tie the FRAND commitment and SEPs availability to a requirement that the licensee agree to license back unrelated SEPs or non-SEPs.

Recommendation 3:4

SSOs should clarify in their policies that a holder of FRAND-encumbered SEPs may require a licensee to grant a license in return under FRAND terms to the SEPs it owns or controls (and those of its affiliates as specified in the SSO's policy) covering the same standard or, as specified by the SSO, related standards.

Recommendation 3:5

It should be understood that SSOs' IPR policies do not affect the freedom of parties to voluntarily enter portfolio or cross licenses beyond the scope of the standard. This includes situations where prospective licensors offer to license SEPs in a package, such as a fixed pool.

4. SEP Disclosure and Information Transparency

4.1 Disclosure as an Element of SSO IPR Policies

Patent disclosure is one of the main elements of many IPR policies, with the objective of increasing the degree of knowledge and transparency relating to patents in the course of the standards development process, and for subsequent licensing. Information on disclosed patents and, often, associated licensing commitments is usually made public in databases published by standard-setting organizations (SSOs). In this chapter, we discuss disclosure rules across a set of SSOs, the roles disclosure can play, and the current level of available information on disclosed patents. This chapter is limited to the disclosure of patents and does not address the disclosure of licensing terms (Bekkers and Updegrove, 2012).

Many SSO IPR policies have disclosure rules as central elements. In fact, such obligations often represent a very significant part of the overall text of these policies. However, there are also SSO IPR policies that do not include disclosure obligations at all. For example, ANSI does not require its accredited standards developers to include such requirements, although it encourages them to do so. There also are many SSOs that have no overall disclosure requirement. Rather, commitments are triggered by participation in the process.

Generally, disclosure rules specify when and how members or participants should inform the SSO that they believe they own patents that might be essential to the standard when it is finalized. However, the exact rules show a great degree of variety. Some important dimensions in this regard include the following:

What triggers a disclosure obligation? Often, it is triggered when participants or members are reasonably aware that they or the companies they work for likely own patents with essential claims based on the then-current drafts of the standard. Such essentiality cannot be definitively known until the standard is finalized. Disclosure obligations are frequently linked to participation in specific working groups or the submission of technical proposals. Although the policies of all surveyed SSOs and of the SSOs of which the committee Is aware, explicitly state that patent searches are not required some state that participants and companies must act in good faith. For instance, a company is not acting in good faith if it purposely (i.e., with the intention of circumventing IPR policy rules) sends delegates who are not personally aware of certain relevant patents to participate at standards-setting meetings.

Whose patents must be disclosed? Almost invariably this obligation relates to all relevant SEPs owned by the individual and the company that person works for. The large majority of policies also require that SEPs owned by affiliate companies (e.g., direct and indirect subsidiaries and sometimes parent and sibling companies) are included. In addition, quite a few policies encourage or require disclosure of known essential patents owned by third parties.

What information must be disclosed? Some SSOs require that disclosures identify all the specific patents or patent applications that are believed to include potentially essential claims. Many, however, allow for blanket disclosures, which are statements that the company believes it owns patents that may end up being SEPs but not listing them.

How is essentiality defined? There are many different attributes that such definitions may incorporate and SSO policies display great variety in the details. However, most SSOs' policies refer just to SEPs that are technologically required in order to comply with the standard. The assessment of essentiality is somewhat subjective and actual determination can only be made by a court.

When must disclosures be made? One common approach is for the policy to allow disclosures at any time during the standards development process. Here, policies typically require timely disclosures, triggered when the representative of a company becomes aware that it may have an essential patent. Few SSOs, however, have exact rules on this. Several SSOs stress that there is a tradeoff between the timeliness and the quality of disclosures. In fact, it can be difficult for companies to project what might end up being essential until the text of the standard is almost finalized. Another common approach is to expect disclosure at particular points in the process, such as 30 to 60 days from the date when a draft is posted. This is usually imposed in addition to the ongoing duty to disclose.

To whom is disclosed information made available (and which information)? Many SSOs make disclosure information available to the general public, but some do not. There is great diversity in the completeness and accessibility of the information that is made public. Rules relating to the recording and availability of disclosure statements made orally during meetings may be unclear. Often such disclosures at a meeting are incomplete and the individual's employer must determine whether a formal statement needs to be submitted to the SSO.

It is important to recognize that because disclosure obligations can create significant costs and burdens for participating patent holders, many SSOs are willing to accept broad licensing commitments in lieu of detailed disclosures.

4.2 The Possible Roles of Information Disclosure

Few IPR policies of SSOs are explicit about what disclosure rules or guidelines aim to achieve. It is possible that many disclosure policies aim to serve multiple goals. The fact that these goals may require different and sometimes conflicting policy elements (e.g., regarding the timing of disclosures) also contributes to the ambiguity of these rules.

The committee's background study of disclosure policies suggests that they may serve at least one or more of the following four distinct goals. The first objective is to allow working group members to make appropriate and informed choices concerning the inclusion of technologies, based on technical merit, implementation costs, and the prospective availability of licenses. Working groups may also use disclosure information to choose between different technical alternatives or to mount efforts to design around a certain patented technology. In the IETF, working group members are known to have frequently considered disclosure information in this respect.[1] A second goal is to record which members and participants are subject to licensing obligations for SEPs following directly from the policy. Third, disclosure serves as a trigger so that essential patent holders can be requested or required to make a related licensing commitment. Finally, disclosure rules inform prospective implementers about which companies they may want to approach to seek licenses and to allow them to assess the extent and value of the claimed patents.

SSOs vary in terms of which goals they try to achieve, depending on the specific context. For a relatively narrow standard, the working group may only face a handful of SEPs and might thus pursue the first goal above. In contrast, an extensive and complicated standardization effort might incur thousands of disclosures, making this much less realistic. Often, SSOs may try to achieve multiple objectives, which might require different and sometimes conflicting elements in disclosure policies, as discussed below.

Stakeholders have diverse interests in disclosure information. The array of stakeholders is long (Raes, 2010). First, working group participants may need disclosed information in order to perform their work. Second, for planning purposes, actual and prospective implementers of the standard may need to know which parties claim to own essential IPR, which specific patents they believe may contain essential claims, whether the IPR holder will require implementers to obtain a license, and if so, whether payment of a royalty or other fee will be required. Sufficiently specific disclosure information also allows implementers to review how many possible SEPs are disclosed, their nature and potential value, and whether implementers agree that the patents in question are valid and essential. Third, SEP owners may use disclosure to assess their claims in the context of those owned by others and develop a general idea of what fee levels might be appropriate within the boundaries of their FRAND commitments.

Policymakers and public authorities are also interested in disclosure. Document disclosure can expand the stock of prior art information available to patent examiners and also be an important input into post-grant opposition proceedings, long established at the EPO and expanded in the United States under

---

[1] While IETF rules allow for the inclusion of technologies for which FRAND commitments were submitted, many working groups have a strong preference only to include unpatented technology or patented technology available at RAND-RF conditions. Thus, many IETF working groups wish to receive disclosure and commitment information as early as possible in order design around certain technologies, if desired.

the America Invents Act. Perhaps most concerned are competition authorities, who may monitor standardization processes to ensure that no unnecessary harm is done to competition. When a case of possible anticompetitive behavior is brought to their attention they might consult relevant patent disclosure databases. Such databases may show whether certain parties fulfilled their commitments and may help authorities assess possible anticompetitive behavior. More generally, policy makers may have an interest in SEP disclosure databases to understand how reliant specific industries are on SEPs.

Courts proceedings may rely on disclosures as matters of record for establishing compliance with the rules of an IPR policy. They may provide key benchmarks of behavior during the standards development process and help determine which parties are bound to specific commitments. Moreover, several courts and competition authorities have embraced the view that FRAND fees should bear a reasonable relationship to the economic value of the IPR prior to its inclusion in the standard (Federal Trade Commission, 2011; European Commission, 2011). Thus, accurate disclosure information is an important potential input in court proceedings.

4.3 Levels of Disclosure

Our survey of actual disclosure information in SSOs suggests that there are limitations to the data and the transparency it provides with regard to the availability, quality, accuracy and comprehensiveness of disclosed SEPs. We next discuss the most important elements that affect the degree of transparency, noting that the underlying choices often reflect a tradeoff among different objectives or concerns.

Under-disclosure and over-disclosure

Study of SSO databases suggests that current lists of disclosed patents display a high level of both under-disclosure and over-disclosure of patents. There are at least two underlying reasons: (1) the incentives that drive organizations into setting certain levels of disclosure, such as the costs or effort of making them and the legal risks of not doing so; and (2) the nature of the IPR rules selected by SSOs.[2]

Under-disclosure refers to a situation in which some SEPs are not present in the IPR disclosure lists. One obvious cause is that IPR policies can only bind members or participants and not third parties. At best, identified third parties can only be requested to provide disclosures and possibly licensing commitments. A requirement that SSO participants disclose third-party patents that might be essential can potentially fill this gap. However, the level of disclosure resulting

---

[2]Industrial economists sometimes argue that over-disclosure in particular could result from the strategic behavior of patent owners, who may attempt to suggest early that they deserve a claim on licensing revenues or attractive cross-licensing conditions. .

from such rules can be both erratic and of limited reliability. Such disclosures may also be restricted for confidentiality or other legal reasons.

A second reason for under-disclosure is that the disclosure obligation, in the context of a specific standard falling under a particular SSO's IPR policy, may not be triggered at all, even if the IPR owners are members or participants. As noted above, disclosure obligations are often linked to actual participation in a working group or to the submission of a technical proposal, and to actual knowledge of patents or patent applications. A company that is an SSO member but does not participate in a particular case may not have an obligation to disclose its IPR. No SSOs in the surveyed group have an all-encompassing patent search requirement. A third factor is that disclosure rules are often subject to the individual knowledge of participants. While there may be additional good-faith requirements, it is possible that member companies will own relevant SEPs and yet the disclosure requirement may not be triggered due to the specific wording of the IPR policy.

Over-disclosure refers to patents listed as possibly essential that end up not being essential in the final version of the standard. A major cause is that companies have strong reasons to disclose and thereby be on the safe side. Several legal cases have held that a company in certain circumstances found to have intentionally failed to disclose was prohibited from commercially exploiting the non-disclosed SEPs later on.[3] Thus, many firms may decide that it is better to disclose too much than too little. A second factor is that over-disclosure takes less time and costs less than undertaking the effort of determining whether specific patents might be essential in the draft standard, especially as its text evolves. Lastly, over-disclosure can be a consequence of the lack of a requirement to update information, as discussed below.

Although over-disclosure may appear less costly than under-disclosure, this is not necessarily clear-cut. As a strategic matter, over-disclosure could reveal to competitors more than firms would like about the patents it believed to be potentially relevant. Under-disclosure has the corresponding strategic advantage of preserving non-public information about patent ownership.

Both over-disclosure and under-disclosure can have market consequences as each may result in legal uncertainty for implementers. For example, if implementers believe there is a substantial degree of under-disclosure, they may decide not to adopt the standard in question because of the legal and financial risks of not being able to assess actual SEP ownership. Similarly, substantial over-disclosure could impede the adoption of standards if implementers are concerned that there may be more patent owners seeking to license more SEPs than there actually are.

---

[3]See, for example, *Qualcomm, Inc. v. Broadcomm, Inc.*, No. 2007-1545 (Fed. Cir. Dec. 1, 2008) and the Dell VESA case, *in re Dell Corporation*, 121 F.T.C. 616 (1996), in which a consent agreement was reached.

Blanket disclosures

SSOs that allow blanket disclosures of patents and patent applications often let the submitter decide whether to make blanket or specific SEP disclosures. Sometimes blanket disclosures are only allowed when certain conditions are met. At IETF, for example, they are only permitted if the owner also commits to licensing its patents on FRAND-RF terms. At ITU blanket disclosures are allowed only if the related licensing declaration does not contain a refusal to offer FRAND or FRAND-RF licenses to ultimately essential patents.

Blanket disclosures entail both advantages and disadvantages with regard to licensing. One advantage for SSOs is that if such disclosures are accompanied by licensing commitments, those commitments will cover any SEP that the submitter has reading on the final version of the standard. In some SSOs, licensing commitments associated with specific disclosures apply only to those particular patents.[4] Further, allowing blanket disclosures may increase the willingness of firms to be SSO members, participate in work programs, and make technical submissions.

Blanket disclosures also provide advantages for SEP holders. They may reduce the legal risks SEP owners might face with under-disclosure. They permit firms to avoid incurring costs associated with specific disclosures. For firms with large patent portfolios, such costs would be both high and recurring. Some of these companies do not routinely seek to monetize their SEPs, so making specific disclosures represents an unnecessary cost. In contrast, companies that do monetize their SEPs see identifying their claims as an investment. Concerns have been raised that forcing companies that do not seek monetization to make specific disclosures may cause them to seek licensing revenues to offset the associated costs.

At the same time, blanket disclosures have a number of disadvantages. First, they make it difficult for engineers to invent around patents. Second, they can shift search costs onto other parties, such as prospective implementers, working group members, or other stakeholders. Third, blanket disclosures may create a situation in which prospective licensees have limited information about the exact magnitude and content of an essential IPR portfolio. In principle, this situation could raise the relative bargaining power of SEP holders that proactively seek licenses and garner better contract terms for them, especially if their negotiation partners cannot readily determine essentiality.[5] Whether this is a prac-

---

[4]In the IPR policies reviewed in the Bekkers-Updegrove paper, this was the case for 4 of 10 policies. In the other six cases the commitment covers any essential claims under the specific standard in question, regardless of whether these patents were actually disclosed by their owner.

[5]This situation describes a case of asymmetric information, which the theoretical literature links to an improved bargaining position for the party with the greater knowledge (Spence,1973, Gallini and Wright, 1990).

tical concern in actual negotiations is unclear, for the parties can ask each other for detailed information.

No requirements to update

There are various reasons why patents or patent applications disclosed as essential at one time may later be deemed as nonessential: (1) the final version of the standard no longer covers the patented technology; (2) the patent application was rejected, successfully opposed, or abandoned; (3) the relevant patents expired; (4) patents with essential claims were successfully challenged in court, or rescinded on reexamination by the relevant patent authority; (5) the scope of the issued patent was narrowed or modified and no longer contains claims that are essential to the standard; and (6) new technical alternatives can arise. Further, even if claims under a disclosed patent end up being essential to the final standard, new information that could be useful to implementers may become available over time.

It should also be noted that SEP ownership often changes hands and a new owner may decide to require payment of a FRAND fee even though the original owner may not have required any fee at all.[6] Finally, there may be complex situations with regard to inventions that are patented in various countries around the world. It is common for the precise scope of patents in such a patent family to differ by country. Thus, a family member may be essential in one country but not in another.

All of the factors above may affect the accuracy and validity of information contained in disclosures and several IPR policies recommend that SEP owners update disclosure information. However, few SSO policies provide guidance on how and when any such updating should occur, let alone impose a requirement to do so. While changes resulting from some causes may be tracked from public sources such as patent offices, many changes may not be.

Costs can arise where patent information is not updated. An implementer may find it difficult to determine from whom it should obtain licenses and which technologies that license should include. Also, implementers may need to reconstruct such information multiple times across several potential licensors. These costs may be borne by each prospective licensee, resulting in considerable duplication of efforts.

The accuracy and validity of information contained in disclosures is also related to the rules regarding their timing. If early disclosure is encouraged or required, the updating problem becomes greater because it is more difficult for companies to project what might end up being essential.

---

[6]Transfers are discussed fully in Chapter 5. Note that the increasing prevalence of patent transfers provides a mixed argument for and against blanket disclosures. On the one hand, the buyer of an undisclosed patent may never realize that she/he has acquired an essential claim and therefore might not assert it. On the other hand, if she/he did discover and assert the SEP there would be a competitive change in the marketplace.

ETSI illustrates that stakeholders sometimes value more accurate information. In that case there has been a process of substantial quality updates to the IPR database, by linking the disclosure data to the European Patent Office database and asking the original submitters to correct data believed to be in error.[7] A varied group of ETSI members donated a significant amount of money to fund this upgrade of the database, suggesting they see benefit in more transparency. However, some companies do not make use of the database.

Limited information on third party IPRs

As mentioned above, SSO IPR policies are not and usually cannot be binding on non-members or non-participants.[8] In the absence of effective rules on third party disclosures, there is an incomplete IPR disclosure database.

Disclosures not made public

Although most SSOs make disclosures public, not all do. Furthermore, many SSOs have a 'dual' disclosure policy. This contains first a disclosure process based on written declarations, often using a predefined form or template and often combined with a licensing commitment declaration. The second arm is an oral disclosure requirement for participants present at meetings. Several of the policies reviewed in the background paper specify that the latter disclosures are to be recorded by the meeting chairperson. But the process may be unclear as to how often disclosures are made and to whom the entailed information is available. Typically, any IPR-related information disclosed at a working group meeting is used by the SSO to initiate contact with the implicated company to see if it agrees on essentiality and, if so, whether it will submit a formal declaration.

Discretionary disclosure policy

Some SSOs do not have any disclosure policy at all. Although ANSI encourages accredited SSOs to include a disclosure policy, it is not a requirement.[9] Other SSOs have a participation-based policy, in which participants agree upfront to be bound by certain licensing commitments for any SEPs they have that end up being essential for the final standard, but no disclosures are required.

We conclude this section by observing that many of the aspects affecting transparency are subject to meaningful tradeoffs. Some of these tradeoffs are

---

[7]See Chapter 7 for further discussion.

[8]There are complexities here, especially with regard to ISO and IEC, which have specific rules that place obligations on those involved in the national accreditation process. See Bekkers and Updegrove (2012) for more details.

[9]An ANSI task force is considering the disclosure issue and its revelance to its IPR policy.

SEP Disclosure and Information Transparency

balanced in more conscious ways than others. Where these have been considered, not all legitimate stakeholders have always been part of that decision process. Finally, for companies, there is an important internal tradeoff to manage with respect to transparency. On the one hand, more transparency can reduce their legal risks and uncertainties and can be beneficial in cases of conflict. On the other hand, achieving transparency through disclosures entails significant efforts and compliance costs for companies.

4.4 The Timing of Disclosures in Relation to Licensing Commitment Procedures

When analyzing the IPR databases of large SSOs, it becomes evident that many initial disclosures are submitted long after the final standard is adopted, even if the SSO policies encourage early disclosure. Obviously, such disclosures come too late to allow working group members to make appropriate choices concerning the inclusion of alternative technologies. Why do companies often disclose so late, even if the rules stress the need for 'timely' disclosure, and why do the rules of any SSO fail to require disclosure prior to final adoption of a standard? We believe that one important explanation lies in the IPR rules and associated procedures themselves. The disclosure rules are often intertwined with the licensing rules, with a single form used for both. Most IPR policies are not clear in whether they prefer early or late disclosure. The entanglement of the disclosure and licensing policies encourages late rather than early disclosure.

The committee suggests that SSOs evaluate whether there is a need to address these issues, in which case SSOs could consider one of two approaches. First, disentangle the disclosure and the licensing commitment processes, for instance by introducing 'early general licensing statements' such as that recently introduced by ETSI. Such statements provide assurances of availability but do not constitute disclosures. They can be made before a company examines the draft specifications or knows it owns essential patents. In fact, disentangling could usefully distinguish between blanket disclosures and blanket licensing commitments, which are different promises and exist in different combinations. Suppose, for example, that a company only discloses a single patent, with an associated blanket FRAND licensing commitment. The latter would also commit it to FRAND licensing for other essential patents it might eventually be found to own but did not disclose.

Alternatively, SSOs might adopt a policy that encourages early disclosures and has an updating requirement. This way, knowledge about essential IPR is "correct" at the time the standard is being developed, allowing for informed decisions, as well as at the time the standard is finalized. In this case, the SSO would have to weigh the benefits this would yield against the burdens this would place on patent holders.

Although such procedures may represent advances for some SSOs, we note that many have evolved requirements that seem to strike a workable balance between timing and information disclosure. Many SSO policies require patent calls at every meeting and participants must choose between binding licensing obligations or disclosing previously withheld essential claims) before final adoption of a standard. Disclosure of essential patents is also often required if the right to charge a FRAND royalty is reserved. This time-tested approach may represent an effective balance of interests for other SSOs to consider.

4.5 Recommendations to SSOs

SSOs should consider several actions to increase the transparency of SEP ownership and licensing.

Recommendation 4:1

SSOs that do not have a policy requiring FRAND licensing commitments from all participants should have a disclosure element as part of their IPR policy.

Recommendation 4:2

SSOs with disclosure policies should articulate their objectives and consider whether they sufficiently serve these objectives. In particular, such SSOs may consider separating patent disclosure from licensing commitments and better define their preferred timing and specificity of disclosures.

Recommendation 4:3

SSOs should make disclosed information available to the public.

Recommendation 4:4

SSOs should to consider measures to increase the quality and accuracy of disclosure data. Such measures might include updating requirements or greater coordination with patent offices.

5. Transfers of Patents with Licensing Commitments

5.1 Introduction

A number of technical standards have encountered instances in which standard-essential patents (SEPs) subject to the owner's licensing assurance have been transferred to a third party. As noted earlier, patents and patent portfolios are becoming more frequently transferred assets. In turn, it will be increasingly important and complex for standard-setting organizations (SSOs) and regulators to address transfer issues in the context of FRAND-encumbered standard-essential patents (SEPs).

Patents may be transferred for several reasons. The patent owner may be no longer active in an area of technology, seek to recognize monetary value for its R&D investment through patent sales, exchange patents or patent portfolios with other entities, be insolvent or in bankruptcy proceedings leading it to sell assets, or may have other purposes. Buyers may acquire patents for any number of reasons including for defensive purposes, financial opportunities in owning patents, or a desire to ensure access for themselves and others to the patented technology, among others.

As SEPs are transferred, the following issues arise. First, does a licensing commitment or assurance made by an SSO participant remain valid for implementers after the SEP is transferred to another party? Second, would specific licensing terms and conditions offered by a prior owner apply to a successor-in-interest? Third, will the new owner be able to enforce the SEP against implementers of the standard without adhering to the fair, reasonable, and non-discriminatory (FRAND) constraints imposed by the prior owner's licensing commitment? Would injunctive relief be available to the transferee and would an antitrust, estoppel, laches, or other defense be available against a transferee? We address these issues in this chapter by considering the interests of the patent transferor, transferee, and future implementers.

Depending on the issue, considerations such as the firm size, market share, prior relation to the transferor, and nature of the transferee's business may be involved. Uncertainty as to patent transferee rights and the expectations of standards implementers can undermine a vibrant standards environment, a risk recognized by competition regulators, a number of SSOs, and other commentators.[1] Some regulatory agencies have "work-in-progress" proposals aimed at minimizing potential competition problems surrounding FRAND-encumbered SEP transfers. Renata Hesse, Deputy Assistant Attorney General of the Antitrust Division, outlines the following considerations in a speech, "Six 'Small' Proposals for SSOs Before Lunch," at the ITU-T Patent Roundtable in Geneva, Switzerland in October 2012:

> ... I would like to identify for you some policy choices that standards bodies could implement which we believe would promote competition among implementers of the standard, potentially benefiting consumers around the world. A standards body could:

> ... Make it clear that licensing commitments made to the standards body are intended to bind both the current patent holder and subsequent purchasers of the patents and that these commitments extend to all implementers of the standard, whether or not they are a member of the standards body...

The Federal Trade Commission (FTC) took a similar position in the *FTC v. N-Data* case.[2] In the wake of that case, some leading SSOs revised their IPR policies to include a provision whereby the SEP holder must take certain steps to bind its successor-in-interest to the former's FRAND licensing commitment.[3]

In assessing this issue it is important to balance the interests of implementers, patent transferors and transferees, and other potential stakeholders. If the pendulum were to swing too far towards implementers, patent holders might be discouraged from joining an SSO, leading to more instances in which SEPs are not being covered by any commitment. Patent holders might also avoid investing and innovating in standards, adversely affecting the technical values of specifications. Moreover, there are concerns that overly restricting the ability to enforce patents can reduce the economic value of SEPs and hamper their transfer.

If the pendulum swings too far in the other direction, in favor of SEPs holders, then firms that have invested in implementing the standard based in part on FRAND licensing commitments could be at risk of patent hold-up, including via injunctive relief, by a new SEP owner not encumbered by the existing li-

---

[1] See EU Horizontal Cooperation Guidelines, Section 285, which strongly recommends "a requirement on all participating IPR holders who provide such a commitment to ensure that any company to which the IPR owner transfers its IPR (including the right to license that IPR) is bound by that commitment, for example through a contractual clause between buyer and seller."

[2] Decision and Order, *In re Negotiated Data Solutions LLC*, FTC File No. 051-0094.

[3] See, for example, the Guidelines for Implementation of the Common Patent Policy of ITU-T/ITU-R/ISO/IEC and the IEEE Patent Policy, 2012.

cense terms or commitment. This would undermine the intended effect of FRAND licensing commitments. There is a further possibility that a SEP holder could transfer the patent to a new owner with the intentions of circumventing the FRAND commitment.

Ideally, SSO policies should clarify the nature of rights and obligations transferred with an SEP in a manner that promotes widespread implementation of standards without creating additional transaction costs that could impede the otherwise efficient transfer of patent rights. To achieve that balance, SSOs need to consider both the legal implications of their IPR policies and their practical effects on different stakeholders.

The next section discusses recent legal cases that involve the status of licensing commitment after transfer and ancillary issues. The third section outlines approaches aimed at continuing FRAND undertakings after a committed SEP is transferred (Kesan and Hayes, 2012).

5.2 Cases Regarding Continuing License Commitments

In one case, National Semiconductor Corp. (NSC) participated in developing the IEEE's fast Ethernet Standard (IEEE 802.3).[4] NSC submitted a letter to the IEEE outlining how it would license all interested implementers for a set fee of $1000 as an incentive to encourage selection of NSC's technology for the standard. SEPs owned by NSC were transferred several times and a downstream owner, Negotiated Data Solutions, or N-Data, advised prospective licensees that it would license the patents on FRAND terms, pursuant to the IEEE policy. The N-Data terms differed from those promised by the original patent holder, resulting in a much higher cost to implement the standard. When the new terms were announced, implementers complained that they were inconsistent with the FRAND licensing commitment. N-Data's employees included former personnel of NSC who were presumably aware of the original licensing terms.

The matter was considered by the FTC, which reviewed the facts and entered into a consent decree with N-Data based on the specific facts of the case. Under the decree, N-Data was bound only to license on FRAND terms but also to honor the licensing conditions originally published by NSC.[5] In a 3-2 decision, the FTC majority found that N-Data's efforts to charge higher rates, after the standard incorporating the NSC technology was widely adopted, in part out of reliance on the $1000 commitment, was an unfair method of competition under Section 5 of the FTC Act.

---

[4]*In re Negotiated Data Solutions LLC*, FTC File No. 051-0094.
[5]No specific finding was made as to whether the royalty requested by N-Data was consistent with some definition of FRAND.

The FTC observed in its public statement

> The Complaint in this matter alleges that N-Data reneged on a prior licensing commitment to a standard-setting body and thereby was able to increase the price of an Ethernet technology used by almost every American consumer who owns a computer… But if N-Data's conduct became the accepted way of doing business, even the most diligent standard-setting organizations would not be able to rely on the good faith assurances of respected companies. The possibility exists that those companies would exit the business, and that their patent portfolios would make their way to others who are less interested in honoring commitments than in exploiting industry lock-in…There is little doubt that N-Data's conduct constitutes an unfair method of competition… We also have no doubt that the type of behavior engaged in by N-Data harms consumers. The process of establishing a standard displaces competition; therefore, bad faith or deceptive behavior that undermines the process may also undermine competition in an entire industry, raise prices to consumers, and reduce choices.

The dissent of FTC Chair Majoras in the *N-Data* case observed that the subsequent patent owners had, after the standard was approved, submitted their own licensing statements to IEEE that did not reflect the $1000 licensing fee NSC had offered to all implementers. (Majoras, 2012).[6] However, IEEE had received and published these new licensing statements without any objections or caveats. Also, as the dissent noted, "from the time National submitted its letter of assurance in 1994 and at least until 2002, some patent holders changed or clarified the terms of their letters of assurance – even after the relevant standard was approved."[7] (Majoras, 2012). This dissent questioned whether there was sufficient evidence to support a requisite intent by N-Data: "Even if N-Data were motivated by a desire to strike a better bargain than National made several years earlier, that alone should not be considered a competition-related offense." The dissent further stated that the FTC's discretion in applying Section 5 should be "bounded with limiting principles," for example, a linkage to antitrust law, which is an issue that is still under debate. Majoras also observed that, while the transfer was pending, during the time NSC's original licensing terms were widely available, many implementers chose to operate as infringers without taking a license, in all likelihood because the stated fee ($1000) was so low as to be less than the related transaction costs of executing a formal license.

In *Rembrandt v Harris,* SEPs owned by AT&T and subject to a FRAND license assurance, were transferred multiple times, ending up with Rembrandt, a

---

[6]Order at http://www.ftc.gov/os/caselist/0510094/080122do.pdf and dissent at http://www.ftc.gov/os/caselist/0510094/080122majoras.pdf.

[7]*Id.*

patent licensing entity.⁸ Although Rembrandt initially questioned whether it was subject to AT&T's commitment, it later acknowledged that it was bound to grant FRAND licenses for patent claims essential to the standard because it had actual or constructive knowledge of the commitment when it acquired the SEPs. The defendant sought a FRAND license and a determination of FRAND terms, although it did not concede that the patent claims were "essential." The lower court initially concluded that a FRAND license must be offered. However, this decision was revoked when a multi-district litigation put the patent's validity at issue and the district court expressed concern over the parties' conduct in the case.

The issue of whether a transferee is bound by the FRAND commitment made by a prior owner has also arisen in connection with the acquisition of patent portfolios that include FRAND-encumbered SEPs. In 2011, insolvent Canadian company Nortel proposed the sale of numerous assets, including approximately 4000 patents, in the bankruptcy proceeding on a "free and clear" basis. A number of companies, together with IEEE, filed objections to the "free and clear" nature of the sale, noting that some of the patents being auctioned were subject to unspecified SSO licensing commitments. Ultimately, a group of companies, including Apple, RIM (now BlackBerry), Microsoft, Sony, and Ericsson, formed an entity they called Rockstar Bidco LP to acquire the patents and agreed to abide by Nortel's standards licensing commitments.⁹

Not long after, Google acquired Motorola Mobility, and its sizable patent portfolio, including numerous FRAND-encumbered SEPs. Google similarly stated that it would agree to be bound by Motorola's commitments. The U.S. Department of Justice examined both the Nortel and Motorola Mobility situations and found that the acquisitions were not likely to substantially lessen competition, recognizing both the business importance of patent sales to transferors and transferees and the importance of standards licensing commitments being respected after FRAND-encumbered SEPs are transferred.¹⁰

---

⁸*Rembrandt Tech. LP. v. Harris Corp.*, 2008 Del. Super. LEXIS 400. Del. Super., 2008.

⁹The recent bankruptcy of Kodak also involved SEPs. In the bankruptcy Sales Agreement, SSO commitments are described as encumbrances to which the sale is subject as follows: "The promises, declarations and commitments granted, made or committed, in each case, in writing by Kodak to standards-setting organizations ("SSOs") concerning any of the Assigned Patents pursuant to the written membership agreements, written by-laws or written policies of SSOs in which Kodak was a participant, in each case solely to the extent that (a) Kodak is required pursuant to such promises, declarations or commitments or applicable non-bankruptcy law to bind the Person to whom Kodak transfers the Assigned Patents to such promises, declarations or commitments, and (b) such promises, declarations or commitments constitute interests in property under applicable U.S. federal bankruptcy Law."

¹⁰As noted in a speech by then acting DOJ Assistant Attorney General Joseph Wayland, "[t]he commitments made by Apple and Microsoft substantially lessened the Anti-

In Europe, Robert Bosch GmbH transferred FRAND-committed patents to IPCOM, a patent-holding company, in circumstances where continuation of the licensing assurance made by Bosch was at issue. IPCOM had been seeking injunctions in various jurisdictions against Nokia, HTC, and Deutsche Telekom with mixed results. In the Nokia dispute, Bosch and Nokia had been negotiating a license for Bosch's FRAND-encumbered SEPs from 2003 to 2007, before Bosch assigned its mobile patents to IPCOM (ipeg, 2011). Ultimately, under the influence of the European Commission's Competition Directorate, IPCOM declared a FRAND commitment in 2009, but some litigation is still pending.

In yet another case, a major DRAM manufacturer, Qimonda, filed for insolvency in Germany. The Qimonda administrator sent letters to the firm's existing cross-license counterparties terminating those agreements, arguing that under German law the administrator is authorized to accept or reject executory contracts in its discretion. In addition, the administrator sought to sell U.S. patents owned by Qimonda, including patents disclosed by Qimonda to the SSO JEDEC with a FRAND license assurance, "free and clear" of encumbrances. The administrator asked the U.S. court not to apply U.S. law allowing patent licensees to reserve their rights, but to apply comity and follow German law, which, it alleged, did not include a license preservation right.

The Qimonda case questioned not only whether ongoing license assurances to SSOs were at risk but also whether existing license agreements might be at risk in a non-U.S. bankruptcy.[11] At the time of this committee's report, after District Court remand, a U.S. bankruptcy court supported maintenance of licensee rights, observing that allowing termination would violate U.S. public policy, which allows IP licensees to preserve their rights, and would not adequately protect the interests of the parties as required by U.S. bankruptcy law. The case is on appeal to the U.S. Court of Appeals for the Fourth Circuit.

A related issue arises when a transferee does not know that patents it acquires may contain SEPs attached to particular standards, may bear FRAND licensing commitments, or may be restricted by SSO policies. In the absence of actual or at least constructive knowledge of prior specific license terms, as in the *N-Data* case, it is not clear whether a transferee is bound to fulfill such terms. Moreover, in the absence of explicit price commitments or generally accepted principles for FRAND royalty determination, it is not clear whether a transferee is bound to terms and conditions that would be FRAND for the transferor, who may have different incentives and business interests than the transferee.

In another situation, the owner of a portfolio of SEPs for a standard might sell them separately to different transferees, which could result in implementers

---

trust Division's concerns about potential anticompetitive use of F/RAND-encumbered standard-essential patents. The Antitrust Division observed that Google's commitments did not provide the same direct confirmation of its F/RAND-encumbered standard essential patent licensing policies."

[11]*In re Qimonda*, 470 B.R. 374 (E.D. Va., 2012).

paying more for the same collection of patents than when the single owner licensed them as an aggregate. These and other such situations may be addressed to some extent by an SSO's policy and the restrictions it places on its FRAND-encumbered SEP holders. However, their resolution likely will also depend on specific facts. Whether an SSO sees these scenarios as commercial considerations for parties to address or as sufficiently critical and likely topics to warrant SSO guidance would be up to the SSO and its members.

A further question worth brief discussion is whether antitrust-related concerns may impact the transfer of a FRAND-encumbered SEP. In the competition area, regulators typically consider a patent holder's market power in determining whether an antitrust violation has occurred. While merely having a patent does not necessarily confer market power,[12] the European Commission's Directorate-General for Competition has held that "[i]t suffices to stress that market power can be conferred by a single SEP" when the standard constitutes a barrier to entry."[13]

The U.S. Department of Justice concurs that SEPs can confer market power, noting that

> In particular, the agencies found that when a standard incorporates patented technology owned by a participant in the standard-setting process and that standard becomes established, switching in some cases becomes difficult and expensive, and that the particular technology may gain market power (Wayland, 2012).

The case of *In re Proxim* involves the intersection of antitrust and bankruptcy.[14] Bankrupt Proxim sought to insulate the purchaser in a bankruptcy sale of patents, including SEPs, from allegations that Proxim violated the antitrust laws by manipulating a standard and seeking non-FRAND royalties. The bankruptcy court approved the sale, rejecting the FTC's objection that "the sale [should] not be free and clear of the Commission's regulatory and enforcement powers."[15] The Bankruptcy Court noted that the sale would not be possible unless the new owners would be allowed to take the patents free from any concerns about Proxim's conduct in standardization activities.

Whether a prior owner's inaction or statements, such as a promise to license on specified terms or a posting that it will not assert SEPs, can attach to a

---

[12]*Illinois Tool Works Inc., et al., Petitioners v. Independent Ink, Inc.*, 547 U.S. 126 (2006).

[13]Case No COMP/M.6381 Google/Motorola Mobility, Regulation (EC) No 139/2004 Merger Procedure (2012). http://ec.europa.eu/competition/mergers/cases/decisions/m6381_20120213_20310_2277480_EN.pdf.

[14]Case No. 05-11639 (pJW) (Bankr. Ct Del 2005).

[15]Case No. 05- 11639 (pJW) (Bankr. Ct Del 2005).

SEP successor-in-interest may also be governed by traditional law on estoppel, laches, and detrimental reliance and would depend on the facts in question.[16] Likewise, whether patent exhaustion or an implied license defense in which circumstances authorize a party to use patented technology without being open to an infringement charge by the transferor, may apply against a SEP successor-in-interest may be governed by the facts and existing case law as well.[17]

5.3 SSO Approaches to Sustaining Licensing Commitments

Some SSOs look beyond traditional contract and competition remedies to continue implementer access to FRAND licenses after a FRAND-encumbered SEP is transferred and to prevent a transferee from asserting SEPs free from the FRAND commitment against implementers. Akin to the notion of a servitude or covenant that "runs with the land," some have considered a licensing commitment that "runs with the patent." For example, it has been proposed that licensing commitments should be interpreted as encumbrances that bind all successors in interest. How that premise will be viewed by a court, and how close the real property analog will be viewed with respect to patents, which are deemed under statute to "have the attributes of personal property" (35 U.S.C. 261) are unknown and speculative (Kesan and Hayes, 2012).[18]

Although the servitude theory is uncertain, it should be noted that a number of courts and national laws have recognized that patent assignments are sub-

---

[16] *A.C. Aukerman Company v. R.L. Chaides Construction Co.*, 960 F.3d 1020, 1032 (Fed. Cir.1992). See also *Radio Systems Corp. v. Lalor, No 2012*-1233 (Fed. Cir. 2013) in which a patent holder sent an alleged infringer a letter and took no action for 4 ½ years during which there was detrimental reliance – equitable estoppel was validly raised by the letter recipient's successor.

[17] See *TransCore LP v. Electronic Transaction Consultants Corporation*, 563 F3d 1271(5th Cir 2009) (covenant not to sue was enforced by successor) and *Rembrandt Data Techs., LP v. AOL, LLC et al.*, Case No. 10-1002 (Fed. Cir. 2011) (patent rights exhausted despite a maze of patent transfers), *Pratt v. Wilcox Mfg. Co.*, 64 F. 589 (N.D. Ill. 1893) (finding that a corporate successor was bound by its predecessor's agreement not to sue another party). See also *Wang Laboratories Inc. v. Mitsubishi Electronics America, Inc.*, 103 F.3d 1571 (Fed. Cir. 1997) (implied license where Wang had entered into an agreement with Mitsubishi to manufacture and then sell Wang-developed SIM cards back to Wang, where Wang did not advise Mitsubishi or JEDEC SSO of Wang's patent applications).

[18] Notwithstanding the novel theory put forward in the Kesan-Hayes presentation, the weight of authority and precedent appears to be against the imposition of servitudes on personal property. See, e.g., Thomas W. Merrill & Henry E. Smith, *Optimal Standardization in the Law of Property: The Numerus Clausus Principle*, 110 Yale L.J. 1, 18 & n.68 (2000) ("American precedent is largely, if not quite exclusively, in accord" with the principle that "one cannot create servitudes in personal property"); and Glen O. Robinson, *Personal Property Servitudes*, 71 U.Chi. L.Rev. 1449, 1445 (2004).

ject to existing licenses, although this is not necessarily the case in all jurisdictions.[19]

Cascading obligation

A common approach, adopted by a number of SSOs, to continuing licensing availability after SEP transfer involves a "cascading" contractual obligation, by which a patent holder making a FRAND or other licensing commitment is required to bind its SEP transferee to the applicable licensing commitment. Some policies may be interpreted as covering an unspecified number of successive transfers because the new owner agrees to be bound by the same commitment under the terms of the relevant SSO's IPR policy. This, in turn, imposes the cascading requirement on those agreeing to the commitment. IEEE provides that the original patent owner's transferee notifies and binds its transferees, meaning that their policy cascades.

The common patent policy of ITU/ISO/IEC further provides for the passing down of the licensing commitment with a SEP transfer:[20]

> In the event a Patent Holder participating in the work of the Organizations assigns or transfers ownership or control of Patents for which the Patent Holder reasonably believes it has made a license undertaking to the ITU/ISO/IEC, the Patent Holder shall make reasonable efforts to notify such assignee or transferee of the existence of such license undertaking. In addition, if the Patent Holder specifically identified patents to ITU/ISO/IEC, then the Patent Holder shall have the assignee or transferee agree to be bound by the same licensing commitment as the Patent Holder. If the Patent Holder did not specifically identify the patents in question to ITU/ISO/IEC, then it shall use reasonable efforts (but without requiring a patent search) to have the assignee or transferee to agree to be so bound. By complying with the above, the Patent Holder has discharged in full all of its obligations and liability with regards to the licensing commitments after the transfer or assignment. This paragraph is not intended to place any duty on the Patent Holder to compel compliance with the licensing commitment by the assignee or transferee after the transfer occurs.

---

[19] See, e.g., *Spindelfabrik Suessen-Schurr Stahlecker & Grill v. Schubert & Salzer Maschinenfabrik AG*, 829 F.2d 1075 (Fed. Cir. 1987) ("the viewpoint of the law in this country [is that] a patent assignee under normal circumstances would be bound as a matter of law by its assignor's prior grant of a license to a third party") (quoting court below). See also German Patent Law Section 15(3) and Japanese Patent Law.

[20] *Guidelines for Implementation of the Common Patent Policy for ITU-T/ITU-R/ISO/IEC* 23/04/02.

Arguably, under these provisions the original SEP holder who binds its successor-in-interest is not responsible for the conduct of that new owner, but there is some uncertainty about this view. Statements within SSO policies that the original participating SEP owner obligates or ensures that future successors will comply with the licensing assurance or that it will provide appropriate provisions to achieve that end, could mean that the transferor is not released from this responsibility. In many instances, SSOs are simply silent on whether there is a discharge of the original patent holder. To address any potential risks in this context, a SEP owner who participates in an SSO might consider, in drafting the SEP transfer agreement, retaining the right to comply with licensing commitments it has made.

Special questions related to cascading obligations may arise where SSOs provide for blanket license assurances and where SEPs may or may not have been specifically disclosed to the SSO. That is, SSO members may commit to licensing, on a FRAND or FRAND-RF basis, all patents that contain essential patent claims, whether or not they have been individually disclosed or declared to the SSO.[21]

The ITU/ISO/IEC common policy specifically addresses unidentified SEPs. Briefly, for those SEPs that are identified, the patent holder binds its transferee to the same licensing commitment for those specific patents that are disclosed. For those that are not identified, an arguably lesser duty applies and the patent holder must use reasonable efforts to bind the transferee to the commitment for any SEPs that are essential to the standard in question. The policy suggests that this is accomplished through a contractual provision in the transfer agreement binding the successor to any licensing commitment the original SEP holder has made for any SEPs being transferred.[22]

An SSO policy featuring the cascading approach may also face the question of how an implementer may seek a FRAND license if there is a break in the chain of commitment downstream. In this case, an implementer should be able to challenge the party that is at fault on one or both of two grounds – first, that he is not complying with the obligation to license the SEP on FRAND terms, or second, that he is not complying with the requirement to bind the successor-in-interest to the FRAND commitment. However, the party at fault may not be readily determined or found. If located, it may be held accountable to either provide a license or compensate the injured party for provable damages. However, once the patent is transferred, that party may not have the right to grant licenses.

One potential problem in locating the current owner of a SEP is that, under current U.S. law, patent transfers need not be recorded and the real owner may

---

[21] While some SSOs allow for blanket assurances, others (like ETSI) focus on declared SEPs.

[22] The relevant provision in the *Guidelines for the Implementation of the Common Patent Policy* currently is under revision to clarify that the holder of a FRAND-encumbered SEP must bind its successor-in-interest and that this obligation will cascade to future owners.

be unidentified. SSOs and others in the standards community concerned with patent transfer should support legislative, regulatory, or other measures to require recordation of patent transfers.[23]

The committee recognizes that some parties have raised concerns about mandatory patent transfer recordation. One question involves the burden and cost such a requirement would impose. In this regard, the USPTO provides a simple "check the box" and "attach the transfer document" electronic filing process. While the fee for USPTO recording has been nominal at $40, a revision to the USPTO fee schedule will eliminate the fee totally for recording a transfer electronically, starting January 1, 2014.[24] Some have suggested that recording might be more convenient if linked to payment of periodic fees to maintain the patent, but that would leave several multi-year periods during which ownership could change hands many times without actual ownership being revealed to standards implementers, thereby undermining transparency.

Others have raised concerns about how much information may be sought in the recordation process. At this time, the committee proposes that only the transferees be identified. In addition, the committee recognizes that it would be helpful for transparency to identify the "real party in interest." This might entail more in-depth questions about corporate structure, however, the committee concludes that the gains in transparency will benefit the standardization process. Hence, the committee supports legislation or regulation, especially in the United States, under which recordation of transferees and real parties in interest will be free of charge and informative about patent ownership. Such policies would require recordation of all assignments to achieve transparency. Like all other mandates, this requirement could have unintended consequences, including encouragement of even more opaque ownership arrangements. To the extent possible, such consequences should be anticipated in drafting the regulation statute.

Some observers have proposed that SSOs take further steps to track transfers, such as establishing their own transfer registry. However, the costs and duplication of efforts in establishing one or more recordation programs and maintaining responsibility for updates and the added burdens placed on members in managing and monitoring SEPs and on SSOs in securing information as SEPs are successively transferred would make such measures difficult to implement.

The committee notes that when SSOs publicly post a membership list it can assist implementers in appreciating which patents might be subject to FRAND commitments and what risks there may be regarding patents needed to implement a standard. In the current context, implementers discovering that a patent has been transferred to an SSO member with a FRAND commitment can be helpful.

---

[23]This issue is discussed further in Chapter 7.
[24]See 4226 Federal Register/Vol. 78, No. 13/Friday, January 18, 2013/Rules and Regulations Table 4 – Patent Fee Changes.

ETSI is one SSO that addresses a possible break in the chain with both a cascading approach, whereby each transferor must seek to contractually bind its successor-in-interest to the FRAND commitment, and also a statement that it is intended that the commitment should run with the patent and therefore bind successive transferees.[25] Competition authorities may take action on a case-by-case basis to protect implementers in this regard if anti-competitive acts are found.

As mentioned above, a bankrupt debtor, who may or may not be a transferor along the chain of ownership, can endeavour to sell a SEP free and clear. Although existing patent licenses can be preserved under U.S. law (11 U.S.C. § 365(n)),[26] the committee knows of no precedent for holding that a licensing assurance qualifies as a patent license. As in the Nortel bankruptcy, transferees may agree or be ordered to be bound by the prior licensing commitments as a condition of acquiring the bankrupt company's FRAND-encumbered SEPs. Bankruptcy is a thorny issue. SSOs should consider how the licensing assurance may be framed to "run with the patent," support legislation or cases to achieve that end, or both.

The cascading approach has been endorsed by competition regulators in numerous speeches and papers. The recent joint DOJ-FTC-DG Competition statement encourages the following "improvements to current IPR policies of SSOs:

> IPR policies should create as strong a commitment as possible to bind future owners of the IPR to any FRAND commitments made to the SSO. Clearly a FRAND commitment that becomes weaker or more vague upon the sale of a patent (or undermines a commitment to effective dispute resolution) will not to be as effective in protecting consumers as one in which all FRAND obligations must be transferred in a sale (Kühn, et al., 2013).

Notification

A second approach to addressing SEP transfers involves a patent holder who has made a FRAND commitment "notifying" its SEP transferee that there is or may be a licensing commitment. In this case it is not clear if and how an implementer might force the transferee to grant a FRAND license. Nonetheless, in the *N-Data* case, where the transferee had knowledge of the circumstances and terms of the licensing commitment, the FTC held the transferee to the commitment and the originally stated terms.

---

[25] ETSI IPR Policy at http://www.etsi.org/images/files/IPR/etsi-ipr-policy.pdf. ETSI is taking a position with similarities to ITU/ISO/IEC and IEEE.

[26] This general U.S. rule has been questioned by a non-U.S. bankrupt party in the Qimonda case. SSOs might consider supporting case decisions or laws that prevent the termination of SEP patent licenses regardless of debtor location.

Non-circumvention

A third approach is for an SSO policy to require members not to circumvent their licensing commitments by the transfer of FRAND-encumbered SEPs.[27] This approach provides some protection for implementers, but its exact scope and effect are not known. Although it may prove to be effective, it may be insufficient if interpreted as requiring a specific intent on the part of the transferor, where most transfers are effected for multiple reasons, or too burdensome if interpreted as raising possible liability for the transferor if any related downstream issue arises.

The foregoing approaches may not exhaust all of the possible solutions, but they have gained some traction among SSOs. They may be more effective for some SSO models than others. In short, SSOs need to consider the many facets of this issue, as the interests of different stakeholders vary as do the laws in different jurisdictions that may affect implementation of the policy.

5.4 Recommendations for SSOs and Policymakers

Transfers of standard-essential IP are an important feature of the high-technology marketplace, whether because firms seek to realize the economic value of their patents through selling them or because SEPs are an asset in bankruptcy proceedings. Such transfers raise complex issues regarding the obligations and rights of transferors, transferees, and existing and potential licensees along what may be a long chain of transactions. Judicial rulings so far provide only partial guidance and there are significant differences in law across countries. Nevertheless, major competition authorities generally see value in binding transferees to original commitments.

The committee agrees that a FRAND commitment should travel with the patent when it is transferred, although there are different means and modalities by which that could happen. Recognizing the complexity of this legal terrain, the committee makes the following recommendations for SSOs and public authorities to advance the proposition that licenses and licensing commitments should travel with the patent to minimize uncertainty and additional transaction costs for licensees.

Recommendation 5:1

Where they have not already done so, SSOs should develop meaningful policies by which successors in interest are bound to whatever licensing com-

---

[27]For example, an SSO policy may address situations where a party might seek to circumvent a FRAND commitment in anticipation of joining the SSO or thereafter. An SSO policy also may just specify generally that a SEP holder cannot take any action that would result in circumvention.

mitment (e.g., FRAND) the SEP owner made to the SSO in question under that organization's IPR policy. This requirement should apply whether SEPs are individually disclosed or are covered by a blanket disclosure. These obligations should cascade through succeeding transfers.

Recommendation 5:2

Legislation, case law, or other legal mechanisms should tie licensing commitments to FRAND-encumbered patents needed to implement SSO standards. This should be done in ways that ensure that the commitment automatically runs with the patents.

Recommendation 5:3

It may be difficult to identify patent transfers, because under current U.S. law they need not be recorded. Accordingly, public recordation with the patent office of transfers of all patents, should, as soon as practicable, be required by legislation or regulation. The committee believes that this approach of recording all patent transfers is a practical and effective way of advancing transparency for transfers of SEPs, which may not always be identified as such. The record should identify the real party in interest.

Recommendation 5:4

Bankruptcy concerns are especially complex and raise uncertainty about consistency of licensing commitments. SSOs should develop guidelines to ensure that the licensing assurances made to them remain with the patent in bankruptcy proceedings and support legislation, if necessary, to the same end.

Recommendation 5:5

Competition authorities and international policy negotiators should, through legislation or regulation, find means to reduce inconsistencies across national legal jurisdictions in patent-transfer issues, including in bankruptcy processes.

6. Injunctive Relief for SEPs Subject to FRAND

6.1 Introduction

Court-ordered injunctions, which remove infringing products from a market, typically for a period of time, are a principal remedy for patent infringement. Indeed, injunctions are intended to deter or stop such infringement. Patent holders are typically granted the right to petition for injunctive relief. Injunctive relief includes exclusion orders awarded by the U.S. International Trade Commission (ITC) as well as court-awarded injunctions. An exclusion order directs U.S. customs authorities to stop the importation of infringing products, thereby barring their entry into the domestic market.

Pursuing, or threatening to pursue, injunctive relief for patent infringement becomes a contentious issue when it involves a standard-essential patent encumbered by a FRAND licensing commitment. When a patent holder makes such a commitment to an SSO pursuant to its IPR policy, it typically provides an assurance that it is prepared to make its SEPs available on FRAND licensing terms and conditions to anyone implementing the standard. FRAND commitments provide assurances to standards implementers, who must unavoidably use technology claimed in SEPs, that reasonable licenses to those rights will be made available. The question then arises as to whether a SEP owner that has stated its willingness to license should be permitted to petition for injunctions or exclusion orders against implementers.

This chapter considers the current debate as to how a FRAND commitment should affect a SEP owner's ability to seek, or threaten to seek, injunctive relief. How does the FRAND commitment operate in conjunction with the statutes and case law relating to injunctive relief generally? And how does competition law affect the analysis?

6.2 Views of Competition Authorities

The Justice Department's Antitrust Division, the U.S. Federal Trade Commission (FTC), and the Competition Directorate General of the European Commission have not only taken a strong interest in this issue but also articulated a common position.[1] Their views result in part from the market power considerations associated with SEPs and the use of injunctive relief by SEP holders, especially when the implementer in question is willing to enter into a FRAND license.

Patent hold-up is more likely where a company uses its SEPs to exclude competitors from the market.[2] As FTC Commissioner Ramirez testified before Congress in 2012, "[a] royalty negotiation that occurs under threat of an injunction or an exclusion order may be weighted heavily in favor of the patent holder in a way that is in tension with the RAND commitment." As the court in *Apple v. Motorola* observed, "…once a patent becomes essential to a standard, the patentee's bargaining power surges because a prospective licensee has no alternative to licensing the patent; he is at the patentee's mercy."[3] Consumers may be harmed by this practice, whether through higher royalty payments passed on to them in the form of higher prices or fewer products on the market as a result of injunctive relief.

The Commissioner for Competition of the European Commission has expressed similar concerns (Alumnia, 2012):

> To build a smartphone one needs thousands of standard-essential patents. The holders of these patents have considerable market power and can effectively hold-up the entire industry with the threat of banning competitors' products from the market through injunctions for patent infringements.
>
> By threatening to use injunctions, these companies can also make demands that their commercial partners would not accept under normal circumstances.
>
> For example, fearing exclusion from the market, companies might be forced to share valuable patented inventions with a competitor or pay excessive royalties which are then passed on to consumers.

There is a consensus among competition authorities that injunctive relief in connection with a FRAND-encumbered SEP should be a remedy of last resort.[4] They have uniformly taken the position that potential licensees who are willing to enter into a license agreement on FRAND terms must have the opportunity to

---

[1] It should be noted that the associated position statements are somewhat case-specific and the policies are works in progress.

[2] The general analysis of hold-up is in Chapter 3 of this report.

[3] Opinion and Order of Judge Richard Posner in *Apple Inc. v. Motorola, Inc.*, June 22, 2012 in the District Court for the Northern District of Illinois, Case No. 1:11-cv-08540.

[4] See FTC Senate Testimony, 2012; DOJ Senate Testimony, 2012; DG Competition Press Release, 2012.

have disputes between the parties resolved before any injunctive relief can be pursued against them.[5]

On the other hand, competition authorities have not taken the position that injunctions on SEPs should be barred in all circumstances. Barring injunctions outright could incentivize some implementers of an industry standard to forgo negotiating a license "…if its worse-case outcome after litigation is to pay the same amount it would have paid earlier for a license." (Department of Justice, 2013; U.S. Patent and Trademark Office, 2013). Some argue that an implementer who is offered a truly FRAND license has incentives to enter into the license instead of incurring litigation and other costs. Others claim that such an implementer may also have interests in delaying negotiation and licensing costs and in having the SEP owner incur expenses in commencing an infringement action.

The competition agencies' position that a FRAND commitment limits, but does not ban, injunctive relief has led to a debate over when injunctions for a FRAND-encumbered patent may be sought. In this debate, implementers generally argue that the costs and uncertainties of litigation provide a strong incentive to accept a license on reasonable terms and conditions instead of filing lawsuits and incurring related costs contesting those terms, especially when the SEP holder always can bring an action based on infringement and asking for monetary relief. SEP holders counter that the benefits of delaying negotiation and licensing costs will often outweigh the expected costs of litigation for a prospective licensee, making the threat of injunctive relief a necessary tool to bring "unwilling" licensees to the bargaining table.

By the end of 2012 there were at least four companies under investigation by one or more of these competition agencies regarding their efforts to seek injunctive relief based on a FRAND-encumbered SEP. The FTC has finalized one consent decree—the *FTC-Bosch Consent Decree*—on April 24, 2013 and proposed another one in *FTC-Google Consent Decree*, 2013.[6] These consent decrees will restrict the SEP owner's ability to seek injunctive relief except for certain specific scenarios where it can be demonstrated that the implementer is not a "willing licensee."

The FTC describes the proposed consent decree in Bosch as follows:

> The FTC alleged that, as a member of SAE International, SPX agreed to abide by SAE rules that require companies to license their SEPs on FRAND terms. However, SPX allegedly reneged on these commitments and pursued injunctions blocking competitors from using the standardized technologies, even though the competitors were willing to

---

[5]See DOJ and PTO, 2013; FTC Apple Brief, 2012; DG Competition Press Release, 2012.

[6]Consent decrees represent a settlement agreement between the FTC and a targeted company.

> license the technology on FRAND terms. The FTC charged that this practice had the tendency of harming competition and undermining the standard setting process...
>
> To address the FTC's concerns about SPX's conduct relating to its existing portfolio of SEPs, the proposed order requires Bosch not to pursue any actions for injunctive relief on these patents and to make them available on a royalty free basis to implementers of the relevant SAE standards in the ACRRR market.... Bosch also has agreed not to seek an injunction against such third parties, unless the third party refuses in writing to license the patent consistent with the letter of assurance, or the third party refuses to abide by FRAND terms as determined by a court or other process agreed to by the parties.[7]

In November 2011, Directorate-General for Competition issued a public statement confirming that it was investigating whether Samsung's enforcement of SEPs violated EU competition laws (Robinson and Torello, 2011). In December 2012, the agency sent a formal complaint to Samsung, stating that Samsung's claims for injunctive relief against a willing licensee on the basis of its SEPs "amounts to an abuse of a dominant position prohibited by EU antitrust rules" (European Commission, 2012):

> Today's Statement of Objections sets out the Commission's preliminary view that under the specific circumstances of this case, where a commitment to license SEPs on FRAND terms has been given by Samsung, and where a potential licensee, in this case Apple, has shown itself to be willing to negotiate a FRAND license for the SEPs, then recourse to injunctions harms competition. Since injunctions generally involve a prohibition of the product infringing the patent being sold, such recourse risks excluding products from the market without justification and may distort licensing negotiations unduly in the SEP-holder's favour. The preliminary view expressed in today's Statement of Objections does not question the availability of injunctive relief for SEP holders outside the specific circumstances present in this case, for example in the case of unwilling licensees.

In addition, on May 6, 2013, DG Competition sent a formal complaint to Motorola Mobility. The complaint argues that the company's attempt to seek and enforce an injunction against Apple in Germany on the basis of its mobile-phone SEPs constitutes an abuse of a dominant position. While noting that recourse to injunctions is a possible remedy for patent infringements, such conduct may be abusive where it involves SEPs and the potential licensee is willing to

---

[7] *In the Matter of Robert Bosch GmbH*, No. C-4377 (Nov. 26, 2012). It is noted that the Bosch matter arises in the context of a specific merger transaction. See http://www.ftc.gov/opa/2012/11/bosch.shtm.

take a FRAND license. The complaint states that "...the Commission considers at this stage that dominant SEP holders should not have recourse to injunctions, which generally involve a prohibition to sell the product infringing the patent, in order to distort licensing negotiations and impose unjustified licensing terms on patent licensees. Such misuse of SEPs could ultimately harm consumers."[8]

These and similar actions remain pending. The committee notes that, to date, there have been few factual showings of prices harming consumers in the smartphone industry or of the actions of SEP owners impacting competition or entry into the market.

Although competition authorities in the United States and Europe have expressed concerns over seeking injunctions for SEPs implemented by willing licensees, their ability to address such problems is often limited to *ex post* actions addressing specific instances of prior conduct. For this and other reasons, the agencies have expressed a preference for SSOs to provide more information *ex ante* as to the effect of a FRAND commitment in order to prevent disputes from arising later. As DOJ representative, Fiona Scott Morton observed in 2012:

> One question that I have been asked is, 'What's so special about standard-essential patents versus other patents?' Standard-essential patents achieve their status through the collective action at the SSOs. Harm can occur when companies come together and bestow market power on each other by agreeing on a common technology. FRAND commitments are designed to reduce occurrences of opportunistic or exploitative conduct in the implementation of standards... If the FRAND commitments are so vague and ill-defined as to have little meaning, then consumers may not realize all the benefits of the standard...

As suggested by the statements quoted above, an important issue in these cases is who is a willing licensee and how that may be determined, an inquiry revolving around different facts patterns. As noted above, some regulatory agencies have taken the position that the prospective licensee has to state that it is willing to enter into a truly FRAND license and that the SEP holder must refrain from seeking injunctive relief, if there are any related disputes between the parties, until such disputes have been fully adjudicated by a court of competent jurisdiction.

---

[8]See the DG Competition press release at http://europa.eu/rapid/press-release_IP-13-406_en.htm. It should be noted that Google offered licenses on terms (including a 2.25% royalty) to which DG Competition did not object during the merger of Motorola Mobility (MM) into Google, because they were the same terms offered previously by MM and therefore "...would not substantially alter current market dynamics" (http://www.justice.gov/opa/pr/2012/February/12-at-210.html). Further, Microsoft did not respond to this offer, instead filing a complaint in a U.S. District Court, which prompted the MM filing in Germany on which an injunction was granted. A U.S. District Court later issued an order preventing MM from enforcing that injunction.

Competition agencies are looking to SSOs as a first line of defense in preventing FRAND abuse. "SSOs that set forth well-defined patent policy rules that minimize ambiguity can effectively promote competition." The DOJ has suggested that "...Standard bodies might want to explore setting guidelines for what constitutes a FRAND rate." In short, "... standards bodies whose members choose to take steps such as these will help the market for the standardized product to work efficiently by lowering costs, increasing transparency and reducing uncertainty—all of which benefit innovation and competition" (Scott-Morton, 2012; Hesse, 2012).[9]

Former chief economists of the DOJ's Antitrust Division and the FTC, together with a DG Competition spokesperson (Kühn, et al., 2013), have suggested that SSOs clarify that

> The FRAND commitment should include a process that SEP owners must follow before they can seek an injunction or exclusion order by the licensor. This process would include specifying what steps must be taken by parties to resolve disputes over a FRAND's rate, validity, essentiality, or infringement before an injunction or an exclusion order may be sought against the licensee. Reducing the ability of licensors to threaten to exclude a product from the market will reduce the ability of the licensor to extract royalties above the FRAND rate and other significant licensing conditions from willing licensees. The essence of the FRAND commitment is that the firm has voluntarily chosen to accept royalties rather than pursue a business model based on exclusion. This suggests that there can be no irreparable harm from the use of the SEP. Limits on the use of injunctions or exclusion orders are therefore appropriate.

6.3 U.S. and European Case Law

Although the committee has not found any statistical data on the frequency of companies seeking injunctions on SEPs, the issue has been highly publicized in the past few years, primarily in the context of smartphones and other electronic devices. As mobile phones evolved from being a device for making telephone calls into what are actually powerful, miniature computers, the number of patented features included in smartphones increased substantially (Scott-Morton, 2012; Yeh, 2012). At the same time, new competitors have entered the smartphones market, each with a different level of R&D investment directed at hardware and software improvements (Chia, 2012).

Patenting, competition, and standards activity and related litigation have all increased in recent years, with cases in both federal district courts and the U.S. International Trade Commission. SEPs have been the subject of some of that litigation, either as the basis of an infringement claim, or as a countersuit or

---

[9] It must be noted that the Hesse and Scott-Morton statements are ideas in progress and were not intended to state formal DOJ positions.

counterclaim. To some degree, infringement complaints based on SEPs may be easier to prove simply because companies build their products to comply with the relevant interoperability standards. In some instances, companies asserting FRAND-encumbered SEPs have sought injunctive relief as part of those claims (Federal Trade Commission, 2011). In this context, SEP owner plaintiffs generally allege that defendants resist accepting FRAND terms or refuse to enter into negotiations for licenses.

Companies defending against infringement actions based on SEPs have argued that seeking such relief violated the SEP owner's earlier assurance to license implementers on FRAND terms. These cases have highlighted that the SEP holders were seeking injunctive relief at the same time that disputes regarding FRAND terms and conditions were being litigated in a different venue. In some cases, furthermore, infringement is being litigated in the International Trade Commission where the only outcome offering relief is an exclusion order, not an award of monetary damages. The Commission is expected to assess FRAND commitments when SEPs are involved, as happened in the recent *Samsung v. Apple* case.[10]

The issue of whether the owner of a SEP can seek injunctive relief has arisen in a number of different venues, as explained below.

United States District Courts

In district court, both injunctions and monetary damages are possible remedies for patent infringement (35 U.S.C. § 283, 35 U.S.C § 284). At one time, district courts generally granted permanent injunctions to patent owners whose patents were found to be valid and infringed. However, following the Supreme Court's decision in *eBay v. MercExchange,* injunctions for patent infringement are no longer virtually automatic, and district courts must apply a test based on four traditional criteria. A plaintiff seeking an injunction is required to show proof of irreparable harm, the inadequacy of money damages, that the remedy is warranted after considering the balance of hardships between the parties, and that the public interest would not be disserved by an injunction.[11]

Following the *eBay* decision, defendants in infringement claims involving SEPs have argued that permanent injunctions should not be available for FRAND-encumbered SEPs. SEP owners already have voluntarily committed to license their SEPs on FRAND terms as a *quid pro quo* for having their patented technology included in the standard. Further, they argue that monetary damages

---

[10]To clarify, the ITC has not evaluated FRAND in the sense of taking a position on what the term means, or whether proposed terms in a negotiation are FRAND. Instead, it argued in this case that FRAND is not a *per se* ban on exclusion orders, and that Apple failed to provide an affirmative defense in the form of a evidence that Samsung violated the commitment. More commentary is provided in a sub-section below. See also Disapproval of the ITC's Ruling in Investigation Number 337-TA-794 (*Samsung v. Apple*) http://www.ustr.gov/sites/default/files/08032013%20Letter_1.pdf.

[11]*eBay v. MercExchange,* L.L.C., 547 U.S. 388 (2006).

always are adequate to address any resulting harm and therefore the *eBay* standard of irreparable harm cannot be satisfied.

Judge Richard Posner, in a recent patent infringement case between Motorola and Apple,[12] agreed with these arguments. Posner wrote:

> To begin with Motorola's injunctive claim, I don't see how, given FRAND, I would be justified in enjoining Apple from infringing the '898 unless Apple refuses to pay a royalty that meets the FRAND requirement. By committing to license its patents on FRAND terms, Motorola committed to license the '898 to anyone willing to pay a FRAND royalty and thus implicitly acknowledged that a royalty is adequate compensation for a license to use that patent. How could it do otherwise? How could it be permitted to enjoin Apple from using an invention that it contends Apple *must* use if it wants to make a cell phone with UMTS telecommunications capability—without which it would not be a cell *phone*."[13]

Moreover, the court denied Motorola's argument that because Apple refused to accept its initial royalty offer it was entitled to seek an injunction. This refusal did not excuse Motorola from meeting its FRAND obligation.

Similarly, Judge James Robart, in the case between Motorola and Microsoft echoed these sentiments.[14] Judge Robart wrote:

> The court is unconvinced by Motorola's argument that it has or will suffer irreparable harm to its goodwill and reputation because a compulsory license agreement would encourage others to infringe Motorola's standard essential patents. This is not the case. The court's prior rulings have made clear that Microsoft, as an implementer of the H.264 Standard, must accept a RAND license to Motorola's standard essential patents. Indeed, Microsoft, or any other implementer, is not free to infringe Motorola's standard essential patents, and were that to occur, this court's ruling with respect to injunctive relief may be different. The nature of Motorola's RAND commitments, however, obligates Motorola to grant RAND licenses to any and all implementers of the H.264 Standard. As the court has explained, in the situation where a standard essential patent holder and an implementer reach an impasse during negotiations of a RAND license, the courthouse may be the only forum to adjudicate the rights of the patentee and the third-party beneficiary of the RAND commitment. Certainly, easily measurable litigation costs to enforce one's rights cannot constitute irreparable harm.

---

[12] Opinion and Order of Judge Richard Posner in *Apple Inc. v. Motorola, Inc.*, June 22, 2012 in the District Court for the Northern District of Illinois, Case No. 1:11-cv-08540.

[13] The court also held that the implementer had no duty to negotiate if the offer was viewed as "not FRAND."

[14] *Microsoft Corp. v. Motorola, Inc.*, (LEXIS 170587, W.D. Wash, Nov. 29, 2012).

Judge Robart's decision barred the SEP owner from seeking an injunction on FRAND-encumbered SEPs in other jurisdictions, pending the court's determination of an appropriate FRAND royalty.[15]

On May 20, 2013, District Judge Whyte issued a similar decision, finding that the defendants breached their FRAND licensing obligations in connection with the IEEE 802.11 standard by failing to offer a license to Realtek, or to negotiate one, before seeking an exclusion order and injunctive relief at the ITC.[16] According to Judge Whyte's decision, "This conduct is a clear attempt to gain leverage in future licensing negotiations and is improper." The court also issued a preliminary injunction against the defendants preventing them from enforcing any such exclusion order or injunction until the court determines FRAND terms and Realtek refuses to accept such terms. Realtek can preserve its right to appeal and maintain any arguments it may have relating to infringement, validity, and the like.

Judge Posner's ruling has been appealed and a number of amicus briefs have been filed in support of each party, including an FTC brief in support of the decision. Critics of the ruling argue that Judge Posner improperly creates a categorical rule prohibiting injunctive relief in the FRAND context.[17] These critics argue that existing law, including the *eBay* decision and Section 337 of the 1930 Tariff Act, provides a framework for determining facts and weighing equities in order to assess whether injunctive relief is appropriate. Within this framework, a FRAND commitment is one fact that may weigh against a finding of "irreparable harm" or "inadequacy of damages as a remedy" and, hence, an *eBay* review would operate against an injunction. However, in other instances, these factors may be weighed differently.[18]

Further complex issues arise in judicial deliberations over the appropriateness of injunctions when FRAND commitments have been made. One is the

---

[15]Similarly, as Judge Posner stated in *Apple Inc. v. Motorola, Inc.*, 2012: "To begin with Motorola's injunctive claim, I don't see how, given FRAND, I would be justified in enjoining Apple from infringing the '898 unless Apple refuses to pay a royalty that meets the FRAND requirement. By committing to license its patents on FRAND terms, Motorola committed to license the '898 to anyone willing to pay a FRAND royalty and thus implicitly acknowledged that a royalty is adequate compensation for a license to use that patent. How could it do otherwise? How could it be permitted to enjoin Apple from using an invention that it contends Apple *must* use if it wants to make a cell phone with UMTS telecommunications capability—without which it would not be a cell *phone*."

[16]*Realtek Semiconductor Corp. v. LSI Corp. and Agere Systems LLC*, Case No. C-12-03451-RMW (N.D. CA May 20, 2013).

[17]See "Brief of Amicus Curiae Qualcomm Incorporated in Support of Reversal" in *Apple Inc. v. Motorola*, 2012.

[18]An earlier example was *CSIRO v. Buffalo*, 492 F. Supp. 2d 600, 2007 U.S. Dist. LEXIS 43832 (ED Texas 2007). In this case, a non-practicing SEP holder (an Australian government scientific research organization) was awarded an injunction by demonstrating that an existing infringement precluded its ability to license its patent because other implementers could thereby be encouraged to infringe.

question of what defenses should be available to an implementer when facing a potential injunction. Some argue that the implementation should be limited to asserting the existence of a FRAND commitment where an injunction is pending, while others argue that the implementer should be able to raise any related claims and defenses (such as validity, enforceability and non-infringement). According to the later view, these defenses may have a direct impact on the FRAND determination, including what would constitute a reasonable royalty rate (as prospective licensees are not required to compensate the patent holder for invalid, non-infringed and non-enforceable patents). Having all of the issues resolved in the same adjudication arguably is the most efficient way to resolve the entire dispute between the parties, as opposed to focusing solely on whether the SEP holder breached its commitment.

Others, more sympathetic to SEP owners, contend that, whether in arbitration or litigation, permitting an open-ended presentation of all factors risks injurious delay. Moreover, there may be concerns about the costs and prejudice to the SEP owner if every issue must be resolved before a determination can be made as to whether it has complied with its FRAND commitment.

International Trade Commission

The issue of the availability of injunctive relief for SEPs also has arisen at the ITC, an independent federal agency. The ITC conducts investigations into allegations of unfair practices in connection with imports in violation of Section 337 of the Tariff Act of 1930, including infringement of intellectual property rights. Under Section 337, the U.S. Customs and Border Protection Agency stops the importation of products that infringe IPR into the United States where the "…effect is (i) to destroy or substantially injure an industry in the United States; (ii) to prevent the establishment of such an industry; or (iii) to restrain or monopolize trade and commerce in the United States…"[19] Thus, a basic criterion is the effect on domestic industry. An order is granted "…unless, after considering the effect of such exclusion upon the public health and welfare, competitive conditions in the United States economy, the production of like or directly competitive articles in the United States, and United States consumers, it finds that such articles should not be excluded from entry."[20]

A SEP owner must satisfy Section 337's "domestic industry" requirement in order for the ITC to adjudicate the dispute. This requirement is fulfilled if a patent holder can show significant investment in plant and equipment, significant employment of labor or capital, or substantial investment in the patent's exploitation, including engineering, research and development, or licensing.[21]

---

[19] 19 U.S.C. § 337(a)(1)(A).
[20] 19 U.S.C. § 337(a)(3)(A)-(C).
[21] Note also that in the ITC context, the identity of a party may affect remedies in transfer cases. The transfer of a SEP from a non-U.S. company to a U.S. company could influence whether the domestic industry requirement to achieve relief is satisfied. Also,

After the ITC makes a finding of patent infringement following an expedited administrative hearing process, potential remedies include a cease-and-desist order and an exclusion order. Unlike in U.S. district court, monetary damages are not available in the ITC for patent infringement. This suggests that the holders of a FRAND-encumbered SEP seek relief at the ITC for purposes other than to seek such monetary compensation from an implementer of a standard. The cease-and-desist order prohibits the defendant from selling or distributing infringing articles from existing inventory inside the United States. The exclusion order bars infringing articles from entering the United States. Plaintiffs in several Section 337 actions have argued that they are entitled to seek an exclusion order because the defendant is infringing their patent(s) and the plaintiff's case otherwise meets all of the ITC's stated requirements. The defendants have responded with arguments that an exclusion order based on a FRAND-encumbered SEP is inconsistent with the public interest. Section 337 allows the ITC to consider a number of "public interest factors" when determining whether to grant an exclusion order or a cease-and-desist order, such as the public health and welfare, competitive conditions in the United States economy, and United States consumers.[22]

Similar to the cases in district court, defendants have argued that a SEP owner's FRAND promise forecloses its ability to seek injunctive relief.[23] The FTC has encouraged the ITC to adopt such a rule, but the ITC has so far not done so. The FTC has urged

> RAND-encumbered SEPs present considerably different issues. A RAND commitment provides evidence that the SEP owner planned to monetize its IP though broad licensing on reasonable terms rather than through exclusive use. Consistent with the proper role of the patent system, remedies that reduce the chance of patent hold-up associated with RAND-encumbered SEPs can encourage innovation by increasing certainty for firms investing in standards-compliant products and complementary technologies. Such remedies may also prevent the price increases associated with patent hold-up without necessarily reducing incentives to innovate.[24]

This issue of injunction orders under infringed FRAND-encumbered patents also is on appeal to the Court of Appeals for the Federal Circuit, an appellate

---

whether a transferee is or should be bound by a FRAND licensing commitment may be a significant factor that the ITC will need to consider under its statutory framework.
[22] 19 U.S.C. §§ 337(d)(1) & (f)(1).
[23] Gaming & Entertainment Consoles Recommended Determination, 2012; Wireless Communication Devices Initial Determination, 2012.
[24] FTC *Wireless Communication Devices* Brief, 2012.

court that hears appeals from ITC actions.[25] Under Section 337, the question may be if or how and when noncompliance with a FRAND commitment creates a public-interest exception, based on adverse effects on competitive conditions or U.S. consumer interests, that can prevail over ITC relief available when there is harm to a domestic industry caused by a patent infringement.

The extent to which such issues are in flux is illustrated by the recent case *InterDigital Communications, Inc., et al v. Huawei Technologies Co., Ltd., et al*, with relevance to both the courts and the ITC.[26] Here, a court did not accelerate a FRAND determination when the ITC was considering an exclusion order. In a proceeding involving InterDigital patents, Huawei unsuccessfully asked the ITC to stay its exclusion order proceeding until the district court determined FRAND (Rizzolo, 2013). The Delaware district court was asked to expedite discovery on the FRAND issue, but refused. In the order, the court denied defendant's motions for expedited discovery and trial on their counterclaims to determine a FRAND royalty for three patents. The court concluded that "The gist of the request is that each Defendant will be harmed if its products are excluded from the U.S. by the ITC, that the ITC cannot set a FRAND rate, and that the Plaintiff will not offer it a FRAND rate although it has an obligation to do so...." The court did not consider it "practicable" to "...race to a partial judgment here so that each defendant will be in a better position in the ITC litigation." The judge denied the motion for expedited discovery and trial.

In another case, Samsung sought an exclusion order in the ITC against Apple based on FRAND-encumbered SEPs relating to the Universal Mobile Telecommunications Standard (UMTS) for 3G mobile cellular systems for networks. The ITC ultimately issued an exclusion order that would prevent Apple from importing certain iPhone and iPad models into the United States, with one Commissioner dissenting on the ground that the infringed patent was subject to a FRAND licensing commitment and public interest considerations weighed against the exclusion of the Apple products. The Commission, among other grounds, found that Apple did not show that a FRAND commitment precludes exclusion orders under Section 337; that Apple failed to "provide a proper legal interpretation of the FRAND declarations at issue"; and that Apple did not show that the claims at issue were SEPs and that a FRAND commitment was warranted and, if it was, that Samsung had failed to comply with the commitment. On August 3, 2013, the Obama Administration through a letter from the U.S. Trade Representative Michael Froman, disapproved the exclusion order stating that "[t]his decision is based on my review of the various policy considerations discussed above as they relate to the effect on competitive conditions in the U.S. economy and the effect on U.S. consumers." This is the first time an

---

[25]Opinion and Order of Judge Richard Posner in *Apple Inc. v. Motorola, Inc.*, June 22, 2012 in the District Court for the Northern District of Illinois, Case No. 1:11-cv-08540.

[26]*InterDigital Communications Inc., et. al. v. Huawei Technologies Co., Ltd., et. al.*, 1-13-cv-00008 (D Del March 14, 2013).

Administration has veto an ITC exclusion order during the past 26 years.[27] The veto letter elaborated,

> I would like to underscore that in any future cases involving SEPs that are subject to voluntary FRAND commitments, the Commission should be certain to (1) to examine thoroughly and carefully on its own initiative the public interest issues presented both at the outset of its proceeding and when determining whether a particular remedy is in the public interest and (2) seek proactively to have the parties develop a comprehensive factual record related to these issues in the proceedings before the Administrative Law Judge and during the formal remedy phase of the investigation before the Commission, including information on the standards essential nature of the patent at issue if contested by the patent holder and the presence or absence of patent hold-up or reverse hold-up. In addition, the Commission should make explicit findings on these issues to the maximum extent possible. I will look for these elements in any future decisions involving FRAND-encumbered SEPs that are presented for policy review.[28]

In short, jurisprudence is developing in this area, as more cases addressing the intersection of FRAND and injunctive relief arise.

European Courts

Similar to the United States, SEP owners have filed lawsuits seeking injunctions for patent infringement in various national courts in EU member states. As in the United States, European court outcomes have not established clear, uniform jurisprudence.

Some courts have been reluctant to grant injunctive relief in connection with FRAND-encumbered SEPs. For example, a 2012 judgment of the Higher Regional Court of Karlsruhe in Germany overturned a ruling of a lower court that had granted an injunction for a SEP on the basis that a request to restrict sales would infringe EU competition law.[29] Similar rulings involving other companies were issued by courts in the United Kingdom and the Netherlands.[30] In

---

[27]The ITC awarded Apple an exclusion order against Samsung for certain of its products, based on non-SEP patents, shortly after the President disapproved the ITC order for Samsung against certain Apple products.

[28]See http://www.ustr.gov/sites/default/files/08032013%20Letter_1.pdf.

[29]*Motorola v. Apple*, 2012, Higher Regional Court of Karlsruhe, Federal Republic of Germany, Case No. 6 U 136/11.

[30](*IPCom v Nokia and HTC* [2012] EWCA Civ 567); *Samsung v. Apple* District Court of The Hague, 20 June 2012, case numbers/docket numbers 400367/HA ZA 11-2212, 400376/HA ZA 11-2213 and 400385/HA ZA 11-2215.

December 2012, a SEP owner withdrew applications for injunctions before the national courts in five EU national jurisdictions.[31]

Nevertheless, injunctions have been granted in some SEP cases, particularly by the national courts in Germany. For example, in May 2012 the Regional Court of Mannheim granted a SEP owner an injunction, concluding that such action did not violate EU antitrust law.[32] In doing so, the Mannheim court considered the application of the "Orange Book" defense and determined it did not apply. The Orange Book defense is established by a ruling of the German Supreme Court which found that an owner of a FRAND-encumbered SEP abuses its dominant position if it refuses to grant a license or seeks injunctive relief when certain conditions are satisfied, including *inter alia* that a potential licensee has made an irrevocable, unconditional and binding royalty offer to the SEP owner to conclude a license agreement and the potential licensee pays this amount to the SEP owner or into escrow.

The Mannheim court's interpretation of the Orange Book test does not require that a potential licensee's monetary offer be based on FRAND. Instead, the Mannheim court has taken the position that a potential licensee must offer a royalty that is just short of being "clearly excessive" before a SEP owner's refusal of the offer becomes abusive. Because there likely is a difference between a "reasonable" royalty and a "clearly excessive" royalty, the Mannheim court's decision arguably raises the bar for potential licensees to successfully challenge a SEP injunction by invoking the Orange Book defense in the German courts.

Reflecting the developing law across different EU Member States, a Dutch case took issue with the criteria for the assessment of the defense established in the German Orange Book decision. In *Philips v. Kassetten (Hague), 2010,* a FRAND commitment by a patent holder was found not to be a defense to patent enforcement, including injunction. In that case, involving CD and DVD technology, the Dutch court found that the German "Orange Book" decision is contrary to Dutch patent law, creates legal uncertainty, and is unnecessary to protect the interests of the defendant. Instead, the court ruled that an implementer should seek a license and, if it is not granted, go to court to seek an interim injunction to preclude suit against it, or a temporary license to the SEP, or damages if the licensee's proposed offer is found to be reasonable.[33]

The inconsistency of the law across the European Union is further underscored by the *Motorola v. Microsoft* case. While the German court issued an injunction, a U.S. District Court in Washington barred Motorola from enforcing

---

[31]"Samsung Drops Injunctions Applications Against Apple," by Vanessa Mock, *The Wall Street Journal*, December 18, 2012. (http://online.wsj.com/article/SB10001424127887324407504578187043081010804.html.

[32]*Motorola v. Microsoft,* 2012, Regional Court of Mannheim, Federal Republic of Germany, Case No. 2 O 240/11.

[33]*Koninklijke Philips Electronics N.V. v. SK Kassetten GmbH & Co.* District Court The Hague, The Netherlands, 17 March 2010, Joint Cases No. 316533/HA ZA 08-2522 and 316535/HA ZA 08-2524.

that injunction pending the court's determination of an appropriate FRAND royalty.[34] The U.S. Court of Appeals for the Ninth Circuit found that the district court did not abuse its discretion and upheld its ruling.[35] In considering this history it is worth noting that in most countries, as in the United States prior to the eBay case, injunctions are granted almost automatically when a valid patent is found to be infringed. Thus, regulatory agencies and implementers are seeking legal means by which the FRAND commitment limits that premise.

Because this case involves one jurisdiction and one set of specific facts, it is unclear what practical effects this decision will have on European injunctions where the SEP owner has a FRAND licensing commitment or, more generally, on how injunctions in one country may be viewed by courts in other countries.[36] It is worth noting in this context that similar ambiguity exists regarding whether courts of various nations feel bound by particular IPR policies of SSOs.

A potentially more definitive case in Europe emerges from the March 2013 order by the Dusseldorf Regional Court referring to the Court of Justice of the European Union (CJEU) five fundamental questions about remedies available to SEP holders where infringement has been found. The case, *Huawei v. ZTE* (no. 4b O 104/12) involves two Chinese electronics companies. The court found that ZTE had infringed patents Huawei had declared essential to the 4G/LTE cellular telecommunications standard. However, in light of the December 2012 European Commission (DG-Competition) Statements of Objection (SO) to Samsung over its attempt to get SEP-based injunctions against Apple, the court decided to ask the CJEU for its determination of appropriate remedies. The case is noteworthy in showing that European courts may differ in their views from the EC. In essence, the CJEU is asked to rule whether the EC's position in its SO or the German view in the *Orange Book* case is more consistent with European Law. The opinion of the CJEU, when it is issued, will be binding on DG-Competition and the courts and competition agencies of all EU member states.[37]

6.4 Industry Views

There are divergent views among firms in the industry on the issue of injunctive relief in connection with FRAND-encumbered SEPs. Some companies have expressed the view that a FRAND licensing commitment precludes the SEP holder from ever seeking injunctive relief. They argue that the commitment reflects an agreement by the SEP holder that reasonable compensation will always suffice and the SEP holder has other remedies to address recalcitrant licensees, for example, by seeking monetary relief and/or similar remedies through

---

[34] *Microsoft Corp. v. Motorola, Inc.*, 2012 U.S. Dist. LEXIS 170587.
[35] *Microsoft v. Motorola, Inc.*, 696 F.3d 872 at 879 (9th Cir. 2012).
[36] A recent court decision in China bears on this issue, as discussed in Chapter 8 of this report.
[37] *Huawei v. ZTE*, 2013, Regional Court of Düsseldorf, Federal Republic of Germany, Case No. 4b O 104/12.

litigation. Companies taking this position observe that an implementer would only resort to litigation, and the related expenses that the implementer would incur, if the SEP holder's offer were not reasonably consistent with FRAND. They also argue that implementers would otherwise be unfairly pressured to accept non-FRAND terms to avoid the threat of injunctive relief or an exclusion order, particularly those authorized by tribunals that arguably cannot fully adjudicate disputes as to whether the SEP holder breached its FRAND commitment and set a FRAND royalty rate.

Other companies have taken the somewhat weaker position that, while injunctive relief should never be sought against a willing licensee, it may be needed in situations involving recalcitrant behavior by licensees. They have argued that any disputes as to whether the SEP holder has offered terms that are consistent with the FRAND obligation, as well as any open issues as to validity, infringement, etc., should be first and fully adjudicated by a court or through an agreed-upon arbitration process. Injunctive relief should only be available if the implementer refuses to comply with any such final decision or otherwise may not be able to be compelled to pay an adjudicated amount, for example, because of bankruptcy, lack of jurisdiction, and the like. These companies also challenge the notion that injunctive relief may be necessary for effective negotiations to occur because the SEP holder can sue for reasonable damages. Thus, injunctive relief should only be available if the licensee refuses to accept the court's determination. And they further argue that the licensor has no reason to accept FRAND terms if the threat of injunctions allows them arguably to seek compensation in excess of that. Consistent with the views of competition regulators noted above, even the threat that injunctive relief may be sought makes it difficult to engage in licensing negotiations on a level playing field.

Still other companies argue that injunctions are necessary to bring the licensee to the table and to incentivize a reasonable royalty for the SEP owner, especially where the willingness of the licensee is at issue. The FRAND commitment may limit the availability of an injunction, so that the force of possible injunctive relief may be lessened in a SEP license negotiation. However, the possibility of injunctive relief if a licensee refuses to negotiate at all, or towards reasonable terms, is necessary for effective negotiation to occur. If the licensee is comfortable that the SEP owner is only entitled to FRAND royalties and no injunctive relief, the licensee may have no reason to discuss terms. All a lawsuit will produce is additional expense for the SEP owner and perhaps the same royalty the infringer would otherwise negotiate. Moreover, if competition authorities may readily impose sanctions on the licensor SEP owner if its opening offer departs from what the FRAND model suggests, licensees may be incentivized not to negotiate and SEP owners will be pressured toward artificially depressed royalty terms and other conditions.

There is also a debate as to when a SEP-holder should be permitted to seek an injunction. Some view any petition for that remedy prior to a full adjudication of FRAND as problematic for two reasons. First, seeking an injunction may distort any bargaining process ongoing between the licensor and licensee. This may happen in part because some tribunals could issue such relief on an expedited basis without fully adjudicating any related FRAND licensing disputes. Second, seeking such relief is not necessary, even in jurisdictions where pleadings for injunctive relief must be made up-front in connection with a FRAND licensing dispute. This is because the court has its own enforcement powers should it grant the SEP holder monetary relief and the implementer fails to abide by that outcome.

On the other hand, some firms argue that access to injunctive relief might be waived if it is not pleaded in FRAND proceedings. This could prohibit SEP owners from even seeking an injunction or an exclusion order and seriously prejudice their economic interests.

In this vein, it is also argued that overly constraining SEP owners may result in fewer innovators participating in a standard's development. They may opt instead to seek higher returns on their R&D investment in other non-standard technologies or avoid participation in the standard development process, which could adversely impact the benefits of standards to the industry and the consumer.

6.5 Recommendations to SSOs, Courts, and Government Agencies

The committee believes that a FRAND commitment limits a licensor's ability to seek injunctive relief, including exclusion orders, and recommends the following steps to help avoid or resolve disputes, prevent anti-competitive conduct, and ensure reasonable compensation to SEP holders whose patents are infringed.

Recommendation 6:1

SSOs active in industries where patent holdup is a concern should clarify their policies regarding the availability of injunctions for FRAND-encumbered SEPs to reflect the following principles:

- Injunctive relief conflicts with a commitment to license SEPs on FRAND terms and conditions should be rare in these cases;
- Injunctive relief may be appropriate when a prospective licensee refuses to participate in or comply with the outcome of an independent adjudication of FRAND licensing terms and conditions; and
- Injunctive relief may be appropriate when a SEP holder has no other recourse to obtain compensation.

The committee could not reach unanimous agreement on appropriate venues for adjudicating FRAND disputes. However, a majority of the committee members endorse the following:

Majority Recommendation 6:2

SSOs should clarify that disputes over proposed FRAND terms and conditions should be adjudicated at a court, agency, arbitration or other tribunal that can assess the economic value of SEPs and award monetary compensation.[38]

The committee also could not reach unanimous agreement on the scope of any limitations that a FRAND commitment might place on SEP holders' rights to seek injunctive relief. However, a majority of the committee members endorse the following recommendation in that regard:

Majority Recommendation 6:3

SSOs should clarify that, before a SEP holder can seek injunctive relief, disputes over proposed FRAND terms and conditions should be adjudicated at a court, agency, arbitration, or other tribunal that allows either party to raise any related claims and defenses (such as validity, enforceability and non-infringement).[39]

---

[38] A minority of committee members endorse this alternative recommendation: Courts, agencies, arbitration bodies or other tribunals (including the USITC) that consider patent essentiality, FRAND determination, or public interest factors should be presented with the facts and render injunctive relief decisions based on existing law, such as the *eBay* decision and/or ITC Section 337.

[39] A minority of committee members endorse this alternative recommendation: SSOs should clarify that a SEP owner that has made an offer and offered to negotiate, with a prospective licensee, a license that will embody FRAND terms should be allowed to include injunctive relief in its pleadings when a FRAND dispute is brought to a court, agency, arbitration, or other tribunal that can consider equities, party conduct, reciprocity, and FRAND factors (including FRAND rates and terms).

7. Patent Office-SSO Information Sharing[1]

7.1 Origins and Scope of Information Sharing

Until recently the standard-setting process has operated largely independently from patent examination and grants. However, as the interplay between standards and patents has increased, along with the number of patents in the technologies comprising ICT, so has recognition that the two systems' functioning and integrity are interdependent. Particular technologies are often both vital to the standards in which they are incorporated and protected by patents. At the same time, implementers who are obliged to license standard-essential patents and often pay royalties for their use have a considerable stake in the quality of issued patents.

In determining whether the subject matter of a patent application is novel, patent office examiners rely on databases of previous patents, publications, and other documents, referred to as prior art.[2] The submissions by participants to standards bodies represent a potentially valuable collection of prior art, consisting of patents, patent applications, and technical specifications. These include finalized standards documents, preliminary and temporary drafts, and other disclosures of technical information to working groups.

These standards-related materials are thought to affect 30 to 40 percent of patent applications in certain ICT fields.[3] Patent offices and standards bodies are considering ways of cooperating to increase the availability to examiners of standards documentation that will improve the examination process. One institu-

---

[1]This chapter relies on symposium presentations by Michel Goudelis, European Patent Office; Dirk Weiler, IPR chairman of the European Telecommunications Standards Institute (ETSI); and George Willingmyre, GTW & Associates. The latter's presentation was commissioned by the Committee and incorporated material from interviews with stakeholders and officials in the United States, Europe, and Japan. See http://sites.nationalacademies.org/PGA/step/IPManagement/PGA_072825.

[2]Under U.S. law, to determine if the technology is novel, the USPTO assesses the difference between the technology claimed in a patent application and the technology available to the public through sale, use, publication, patenting, or other means of dissemination.

[3]Committee consulation with an ETSI representative.

113

tion in particular, the European Patent Office (EPO), has concluded groundbreaking agreements with three SSOs to share such information.

The EPO is the regional patent office established by the 1973 European Patent Convention among 38 member states. It examines patent applications submitted by inventors worldwide. Applications approved by the EPO may be granted EPO patents and also granted by the patent offices of individual member states.[4] The EPO processes the third largest volume of patent applications in the world after the Chinese State Intellectual Property Office (SIPO) and the USPTO.

In recent years the EPO concluded memoranda of understanding with three standards organizations. The first agreement was the Institute of Electrical and Electronics Engineers Standards Association (IEEE-SA), which develops global standards for a wide range of IT products and services. The second memorandum was with the European Telecommunications Standards Institute (ETSI), which also produces globally-applicable ICT standards. Finally, the third agreement was with the International Telecommunications Union (ITU), the United Nations specialized agency for creating global standards in information technology and communications.

The three agreements have several common elements: 1) exchange of information and documentation of mutual interest in the field of standards for the benefit of prior art search; 2) collaboration on documentation format definition and dissemination policies to align them with the EPO prior art search needs; 3) contributions to education activities in the field of standards; and 4) self-funding of expenses associated with the agreements.

Of the three arrangements, the 2009 ETSI-EPO memorandum of understanding has created the most robust relationship although the two institutions had previously developed a mutual understanding since EPO became a member of ETSI in 2003.[5] ETSI is a leading body for globally applicable standards for telecommunications and home of world-class standards such as GSM, TETRA, and DVB. Its membership consists of 766 companies and organizations from 63 countries. Its IPR database contains information on those patents and applications notified to ETSI as being essential or potentially essential to ETSI standards. The value of this database arises from both its comprehensiveness and the structure of relationships in its information architecture, in particular its integration of patent documentation, bibliographic information, patent families, and patent number normalization. Under the agreement with ETSI the EPO has acquired documentation on standard-essential patents from ETSI along with necessary bibliographic data, incorporated the data into its internal databases, and

---

[4]In December 2012, 25 European states agreed to pave the way for a unitary European patent issued by the EPO and for the creation of a unitary European court to handle patent disputes.

[5]EPO has joined other SSOs also to access prior art, often under the same conditions and costs as industry members who profit from their membership, but membership is not a *sine qua non* of EPO information sharing with SSOs. See discussion below.

educated its examiner corps on how to use data in assessing prior art (Goudelis, 2012).

The EPO is seeking similar working relationships with other standards organizations. In 2012, the Office extended invitations to enter into similar memoranda of understanding (MOUs) to ISO and IEC. The EPO has also shared what it considers its positive experiences with other patent offices (Goudelis, 2012).

7.2 Benefits and Costs of Information Sharing

One measure of the utility of these standards-related information sharing arrangements is the number of times ETSI standards documents have been cited in EPO patent examinations. These citations show a marked decline over the period of cooperation, from 2000 in 2004 to 884 in 2008. No analysis is available of the number of patent applications rejected on the grounds of prior art from these sources, but one can conclude that the data represent a significant addition to the knowledge base available to examiners in the fields covered and have been used to limit the scope of approved patent claims.

The costs of participating in the information sharing arrangements are non-trivial. For the EPO, the costs include membership fees and for the EPO and SSOs, acquisition and conversion costs. Nevertheless, for the EPO, the arrangements contribute to a reputation for relatively efficient search and high quality examination and the benefits are seen to extend beyond generating patents of higher quality. Improved transparency in the linkages between IPR and standards is seen as a benefit by both SSO members and EPO examiners. Disclosure of SEPs, together with a commitment of FRAND licensing, is a requirement of many SSOs. Nevertheless, the amount and quality of information about declared patents may be less than desired by some users of their databases. Generally speaking, SSOs do not perform checks on the essentiality, validity, and completeness of the disclosures. Where disclosure is not compulsory, SSOs are not in a position to provide assurances about the completeness and timeliness of submissions to their databases.

The interlinking of EPO and SSO databases benefits SSO members by providing updated information on patent applications and claims and generating automatic identification of classifications and types of patent families, all subject to the agency's quality controls. ETSI independently modified its IPR policy to extend disclosure and FRAND licensing commitments from a specified member of a patent family to all existing and future essential patents of that family unless there is an explicit exclusion of specified patents at the time of the undertaking.[6]

ETSI also took advantage of its cooperation with EPO to launch a new information architecture in 2011, increasing transparency, functionality, and user-

---

[6]ETSI has a unique definition of patent families with respect to licensing commitments. This term should not be confused with the more common concept of patent families at the EPO and other patent offices.

friendliness by such improvements as reliable reporting of query results, information on ownership and essentiality status, and links to information on licensing conditions. The estimated cost of these upgrades was 1 million Euros, or approximately U.S. $1.3 million.

A possible future benefit of EPO-SSO cooperation is in tracking changes in patent ownership. Both participants in standardization development and standards implementers have a strong interest in the availability of accurate, updated information about who owns which SEPs, in which patent jurisdictions, and for how long. Also important is information about whether titles have been reassigned and to whom. The EPO's ability to track ownership and assignments extends only through the time at which the patent is granted and the nine-month period following, during which an opposition may be filed. Beyond those times information on transfers resides, if at all, in national patent offices. One suggestion to address this challenge of tracking patent transfers is the creation of Internet-based patent registers maintained by standardization bodies in cooperation with patent offices. This idea continues to be discussed in various forums.

Establishing mechanisms to record patent transfers, preferably in the patent office where the national patent was granted, would help in achieving the goal of identifying the ownership of SEP patents.[7] Despite practical concerns, such measures would enable an implementer or standards developer to search official records to determine who owns a SEP, whether it is declared or not, and whether it has been assigned.

7.3 Legal Status of Standards Information

Final standards are part of the available prior art, except in the case of private standard consortia that do not publish them but make them available only to specific parties under non-disclosure agreements. Further, preparatory documents are treated like other written or oral disclosures, meaning that to qualify as prior art they must have been made available to the public prior to the patent filing or priority date without the disclosure being subject to a requirement of confidentiality. Thus, the prior art standing of a standards draft may be subject to the rules or norms of the SSO concerned and is not always clear, nor is the date of a document always verifiable.

The incentives for members of SSOs to make early specifications available as prior art are often mixed. In some SSOs, circulation of early drafts of specifications is limited to those working on it, perhaps due to a concern that allowing other parties to comment adversely on preliminary drafts could chill

---

[7] Japan records transfers of all patents, including SEPs. An article by Nahoko Ono, *Avoiding Japanization: Lessons from Japanese Gridlock on the Patent Recordation System* (2012) discusses problems with the Japanese system of recording patent documents. The focus of that article is on licensor and licensee concerns with disclosing confidential terms (such as exclusive license scope) and on limiting information to specific information such as grantor, grantee, addresses, and patent numbers.

innovative proposals. There may also be a concern that competing non-members, able to read the new specification before it is formally published as a standard, could somehow undermine it. Such considerations are balanced, however, against an interest in early publication in order to foreclose others from patenting technology created during the standards development process.

Under the three MOUs with SSOs, documents provided to the EPO, whether preliminary or final, are considered not to be confidential unless otherwise specified. To date there have been no cases where the participating SSO has excluded use of shared information as prior art. If a patent applicant were to contest use of such documents as prior art, the circumstances would be assessed case-by-case in the course of patent examination.

7.4 Relevance of the European Experience to the USPTO

The committee considered the implications of the EPO-SSO agreements for the U.S. Patent and Trademark Office, which does not participate in them. In this regard, the Scientific and Technical Information Center (STIC), a central library facility operated by the USPTO, frequently receives requests from examiners to provide the text of standards thought relevant to the patentability of inventions under consideration. The application may reference a particular standard or include part of a standard. Further, the examiner may find mention of a standard in the course of the examination.

The STIC provides access to standards documents through a variety of channels, including most frequently its non-patent literature (NPL) website and its subscriptions to the publicly available standards of some SSOs, such as IEEE-SA. In general, these sources are limited to final standards, and obtaining additional documents relating to them may entail a significant cost to the patent office.

For their part, SSOs have access through public search facilities to the same databases of granted patents and pre-grant applications that are available to examiners. However, like other members of the public, they do not have access to the STIC NPL Website. As for patent transfers, assignments, and re-assignments, the USPTO maintains an assignments database. Submissions of information to it are voluntary, although registration does convey some legal protection that would not otherwise exist.

The committee finds that arrangements along the lines of the EPO-SO memoranda could significantly benefit both the USPTO examination process and SSO functioning at relatively modest cost to both parties. We note, however, two issues that would need to be addressed.

First, as does Europe, the United States has its own jurisprudence on when documentation is "publicly available" and therefore eligible as prior art, and cooperative arrangements must take those laws into account. In a recent pertinent case, for example, *SRI International v. Internet Security Systems Inc.,* a number of patents related to the Internet were alleged to be invalid in view of a prior technical paper. However, the Federal Circuit Court of Appeals ruled that there was "insufficient evidence to show that the paper was publicly accessible as a printed publication as required under U.S. patent law in 35 U.S. Code § 102(b). The paper had been placed on a server, but solely to facilitate peer review and it could not be accessed by the researching public.[8] Holding that the paper was therefore not prior art, the court vacated a summary judgment of invalidity against the patents.

A factor that may simplify and facilitate activity with SSOs is that U.S. patent law has recently changed. In switching the United States from a first-to-invent to a first-inventor-to-file priority system for patent applications filed on or after March 16, 2013, the America Invents Act of 2011 has ostensibly moved U.S. patent law in the direction of European law and other international patent systems. This should make published standards documents in one jurisdiction more translatable to use in other jurisdictions than in the past. Further actions toward harmonizing the requirements for prior art standards publications might be useful, although the process of harmonizing substantive patent laws is inherently difficult and often protracted.

A second issue raised in discussions of prospects for USPTO-SSO cooperation is whether it is appropriate for USPTO to take membership in private organizations, an element of the EPO's relationship with the ITU and ETSI, but not IEEE-SA. Although a number of federal government agencies participate as members in organizations such as the American National Standards Institute (ANSI), the USPTO no longer participates, considering an arm's-length relationship to be the appropriate way for a regulatory agency to support the private sector standards development system while avoiding potential conflicts of interest. Membership may contribute to a certain level of trust in the EPO's relationships with ITU and ETSI, but it is the process of arriving at the MOUs, benefiting from them, and maintaining them that fosters a belief that they are strategic assets.

Along this line, it should also be recognized that U.S. federal and state governments, like governments in other countries,[9] play a significant role as a purchaser of standardized products (DeNardis, 2012). Accordingly, government agencies have an interest in robust, accessible standards that promote interoperability, cost-efficiency, innovation, transparency in the inclusion of technologies in standards, and a diverse system in which no single company controls the standard or is its sole implementer.[10] Cooperation between the USPTO and SSOs could contribute to these objectives.

---

[8] *SRI International Inc v. Internet Security Systems Inc.*, 511 F.3d 1186 (Fed. Cir. 2008).
[9] See also Chapter 8 regarding India and e-government.
[10] Office of Management and Budget OMB Circular A119 directs the government to adopt commercial standards rather than derive unique government specifications unless there is justification.

7.5 Recommendations to the USPTO and SSOs

The committee finds that the EPO's information sharing arrangements with leading international SSOs have demonstrated their value to both parties of having broad, timely, and low-cost access to standards prior art with important potential gains in transparency and efficiency. Similar arrangements could achieve such gains also for the United States even if U.S. law more narrowly circumscribes the range of standards documentation that qualifies as prior art. It should be possible for leading SSOs to agree on terms of cooperation with the USPTO that enable mutually beneficial information sharing without raising concerns about conflicts of interest.

Recommendation 7:1

First, in the wake of the passage and implementation of the America Invents Act, the USPTO should

- Clarify how the legal definition of prior art varies across jurisdictions, particularly as between the EPO and USPTO. Specifically, when is art "publicly available" in a standards context?
- Explore with leading SSOs, including possibly ETSI, IEEE-SA and ITU, information sharing arrangements similar to those concluded by the EPO;
- Work with other patent offices to establish uniform fields and templates for standards-based prior art documents, such as early drafts of specifications, published minutes, and the like, and deliberate with other offices on the definition of sharable information in this context;
- Improve standards technology education for U.S. patent examiners. For example, when standards developers convene in Washington, they could be asked to instruct and update USPTO examiners about standards processes and recent developments; and
- Develop joint education programs with SSOs on the pros and cons of standards-based prior art, especially early drafts, and benefits from including it in patent office search databases.

8. IPR and Standards in Emerging Economies

8.1 Introduction

As part of its charge, the committee was asked to consider the ways in which intellectual property issues affecting standardization are handled in countries and regions other than to the United States and Europe, the main subjects of this report. In further consultation with the U.S. Patent and Trademark Office, we were asked to focus on three major emerging economies: China, India, and Brazil. Presentations and discussion at the October 2012 symposium made it clear that questions of patents in standards are only beginning to attract attention in India and Brazil. In China, on the other hand, where there has been an explosion of patenting and an active standards development strategy, the issue has received considerably more attention, although China has yet to settle on clear policy directions (Breznitz and Murphree, 2012; Ramakrishna, et al., 2012; Barbosa, 2012).

In all three cases, the economies are growing rapidly, are very large in terms of absolute scale and geographic scope, and all have increasingly important roles in global production networks. Each has a large and growing technical community of scientists and engineers, and a research and development system of growing sophistication but only recently have these begun to have an impact on the innovation capacities of the three economies. Instead, there has been a heavy reliance on imported technologies. But in all three countries we now see national governments making significant new industrial policy commitments intended to foster national innovation capabilities and push their economies up the value chain to higher value-added, more knowledge-intensive production. This is especially true in China where growth in R&D expenditures and patenting activity has outstripped Indian and Brazilian initiatives (Adams, et. al, 2013). The approaches in all three to IPR and standards development should be seen in the context of these broader industrial policy goals, including reforms intended to shape the R&D systems to more fully serve technological innovation.

Unlike Japan's industrial development and to a lesser extent South Korea's, where standards development was seen by many observers as strictly a tool of industrial policy, the recent development experiences of China, India, and Brazil are conditioned by their membership in the WTO and the necessary commitments to the Agreement on Technical Barriers to Trade (TBT) and the Agreement on Trade-Related Aspects of Intellectual Property Rights (TRIPS). These obligations require them to attempt to harmonize their IPR and standards regimes with international norms while also opening their economies to greater foreign trade and investment.

Arguably foreign investment has contributed significantly to the growth and modernization of these economies, but it has also brought home to them the importance of IPR and standards for shaping the relative and absolute gains from engagements with the international economy. While the *absolute* gains are readily appreciated, from a *relative* gains perspective, the greater share of the benefits accrues to the owners of the intellectual property and to those with control over standards. All three countries have been challenged, therefore, to build capabilities for intellectual property creation and standards development in order to alter the terms of the relative gains and capture more value from engagement with the global economy. This objective has inevitably led to questions about the respective roles of the state and industrial enterprises in building those capabilities, and about the compatibilities of national technological development efforts with international, especially WTO obligations.

The multilateral pressures on emerging economies to conform with international standards, or to manage domestic policy changes subject to global constraints, may become even stronger as countries join various bilateral and regional agreements. Such agreements increasingly embody stronger requirements for intellectual property protection because IP owners in the United States, EU and other advanced nations generally see the WTO rules as inadequate (Maskus, 2012). For example, although the contents of ongoing negotiations remain confidential, indications are that the Trans-Pacific Partnership (TPP), a U.S.-led agreement among 12 Pacific Rim countries at various levels of economic development, will contain rigorous rules constraining the ability of national governments to limit the scope of patent rights. China has not formally joined the TPP talks but is considering its interests in light of its potential exclusion from a major regional trade grouping. Thus, if China joins there may be additional constraints on its policy freedom in the area of patent use in technical standards. In the discussion below, we focus first on the Chinese case, returning to the Indian and Brazilian experiences at the end.

8.2 China

Although China has a long history of standardization, it is only recently that standards and intellectual property have figured prominently in national policy.[1] This prominence derives from China's accession to the WTO in 2001

---

[1] China joined the ITU in 1920 and rejoined the ISO in 1978. China's Standardization Law was promulgated in 1989 well before developments in the ICT industries changed the landscape for standard-setting and brought issues of intellectual property to the fore.

and the introduction of standards and IP concerns into its national technological development strategies. In joining the WTO, China committed itself to attempt to harmonize many of its standards with international standards and, in the process of doing so, initiated reforms of its standardization bodies. Thus, the current institutional arrangements, in which the Standards Administration of China (SAC) has responsibility for national standardization within the General Administration of Quality Supervision, Inspection and Quarantine (AQSIQ), dates only from 2001 (see Figure 8-1).

Similarly, the national policy focus on standards and IP development as tools of industrial policy is also a largely post-WTO phenomenon, part of China's efforts to put together the pieces of a national strategy for technological innovation. Although the Chinese regime had practiced active science and industrial policies since the 1950s, these gradually began to take new forms after 1980 as China faced expanding engagement with the international economy and initiated a series of domestic economic reforms. By the late 1990s, China was attempting to create an innovation system suitable for a market economy and for engagement with globalization, a challenge which became more pressing with WTO accession. China's experience from the two decades of reform and engagement with the international economy after 1980 led it to conclude that control over standards, and the intellectual property contained therein, conferred considerable market power, while being solely an IPR-poor "standards taker" put it at a considerable economic disadvantage especially with regard to royalty payments. Hence, major new initiatives were begun after 2000 to develop national standards and IP strategies (Ernst, 2011; Suttmeier and Yao, 2011).

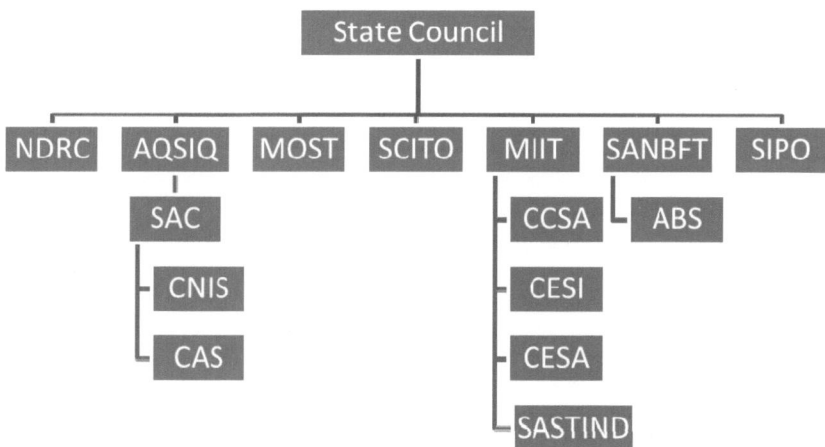

FIGURE 8-1 Central Government agencies for standardization. Source: Adapted from USITO, 2007.

Unlike India and Brazil China displays a more strategic sense of the importance of standards and the incorporation of its own IPR into standards. But, like India and Brazil, it is a latecomer in building a national regime for standards and IP in high technology industries and in learning to negotiate the differentiated terrain characteristic of international standards and IPR regimes.[2] Chinese firms vary greatly in their degree of integration into the evolving legal framework and their sophistication in managing intellectual property in a standards context. China thus strives to learn lessons from the international system while at the same time attempting to develop national industrial policies incorporating standards and IPR strategies as domestic policy tools.

The Chinese government has also attempted to reconcile a strong belief that the government should play a leading role in standardization, with a recognition that commercial entities responding to market forces and rapid technological change play critical roles in shaping standards in the ICT industries. In striving for institutional reconciliation of markets and government, Chinese approaches to standards have also sought to find the proper balance among protecting the interests of private parties who have made investments in technological development and intellectual property, establishing standards on sound technological foundations, and achieving some sort of equity in the distribution of the benefits of standards development. In developing its standards regime, the judgments reached on how to reconcile state versus market roles in standardization and achieve that balance among competing needs has not always accorded with the values and experience of foreign governments and companies, and have, at times, left China in an uncomfortable position *vis–a-vis* foreign stakeholders in established international standardization and IPR regimes (An, 2012).

Key institutions

SAC has broad policy responsibility for national standards. It is supported in its role by a research arm, China National Institute for Standardization (CNIS), and has a close relationship with the China Association for Standardization (CAS), a nominally non-governmental organization for the dissemination of information on standards, provision of certification, and training of standards officials. Nevertheless, industry-level standards are overseen by other entities, the most important

---

[2]The Chinese government has shown a preference for working through the formal international standards development organizations that involve government representation. It is a long-time member of ITU, having joined in 1920, and rejoined ISO in 1978. It has been less comfortable with the workings of non-governmental standards organizations, although this is beginning to change as it realizes their importance and as Chinese firms and China's scientists and engineers become more active in standards development work. For instance, China is a founding member of the 3GPP and participates in the M2M network, and Chinese experts participate in IEEE, NEMA, and other SSOs.

of which is the Ministry of Industry and Information Technology (MIIT).[3] MIIT has its origins in the former Ministries of Posts and Telecommunications and Electronics, which were merged in 1998 to form the Ministry of the Information Industry (MII). In a subsequent government reorganization in 2008, the Ministry's portfolio expanded to include other industries and was renamed MIIT. Although it is active in standardization for other industries as well, it remains a central player in the development of ICT standards, the area of our concern, and the one which has attracted the greatest international attention.

Two key entities under MIIT dealing with ICT standards are the China Electronic Standardization Institute (CESI), first established in 1963, and the China Communications Standards Association (CCSA), established in 2002. CESI, as a former research Institute under the Ministry of the Electronics Industry and the more institutionalized body, has its own research facilities, serves as the secretariat for 11 national technical standardization committees, and works with the nominally non-governmental China Electronic Standards Association (CESA).

CCSA, also a nominally non-governmental associational entity but hosted by MIIT's China Academy of Telecommunications Research (CATR), is governed by a council of leading government, industry, and academic figures in the world of Chinese telecommunications. It maintains a Technical Expert Advisory Committee and a Technical Management Committee, both of which draw on expertise from industry, government, and universities. CCSA also operates through a series of technical committees, under which are working groups and sub-working groups. Although nominally specialized into separate communications and information technology domains, the work of CESI and CCSA at times overlaps and, in keeping with inherited traditions of competition between the Ministries of Electronics and Telecommunications, they have been known to compete with one another.

As shown in Figure 1, a number of other national level bureaucratic entities are also involved in standardization work. These include the National Development and Reform Commission (NDRC), Ministry of Science and Technology (MOST), State Council Informatization Office (SCITO), State Administration of News, Broadcasting, Film and Television (SANBFT) (formerly the State Administration for Radio, Film and Television, or SARFT), State Administration for Science, Technology and Industry for National Defense (SASTIND), and Ministry of Public Security. NDRC and MOST play important roles in promoting China's innovation strategy through subsidies, R&D support, and other policy preferences for standardization and intellectual property development work. SASTIND plays an important role in promoting standards development in military related industries, and SCITO and the Ministry of Public Security have been important in areas of information security standards. SANBFT, supported by a research arm, the

---

[3]China's standards regime provides for four levels of standards: national, industry, regional, and enterprise.

Academy of Broadcasting Science (ABS), has played a key role in the development of broadcasting standards and has figured prominently in Chinese efforts to promote convergence in communications networks, often in conflict with MIIT and its subordinate units. Not included in Figure 1 is the State Administration for Industry and Commerce (SAIC), which has a role in administering China's Anti-Monopoly Law (AML), discussed further below.

Not surprisingly, bureaucratic competition among these national level agencies has, at times, made coordinated national efforts on standardization difficult.[4] When the standards interests of Chinese and foreign companies and Chinese local governments are also factored into the mix, it is clear that reaching common purpose on standards in China is a challenging task.

IPR policies for standardization work

The importance of IPR issues in standardization began to gain prominence in China after 2000. In 2003, MII introduced a draft policy which attempted to clarify the treatment of patents in standards. The policy did not distinguish between essential and nonessential patents, provided for compulsory patent pool participation in the case of mandatory standards, and imposed on foreign companies' obligations which had the effect of privileging Chinese companies. Not surprisingly, it was not well received by foreign stakeholders operating in China, who were given only limited access to Chinese standardization working groups.

Meanwhile, the Audio Video Coding Standard Working Group of China (AVS) attempted to draft a patent policy drawing on what was perceived to be best international practice and which involved participation by foreign companies. In doing so, it solicited input from foreign companies and recruited a leader of the MPEG-4 IPR subcommittee, to chair its IPR working group.[5] The AVS approach was recognized as a positive step forward by foreign stakeholders as well as many in the Chinese standards community who viewed it as an innovative Chinese response to problems plaguing SSOs elsewhere. Its key provisions require members of the working group to sign an agreement consenting to the AVS IPR policy. Members are required to disclose known related patents and to

---

[4]In one widely noted case, for instance, Chinese efforts to develop its own audiovisual standard (AVS) as an alternative to the MPEG standard were set back by SARFT's decision to adopt the latter. This decision was subsequently modified in 2012 when, as a result of improved government coordination, SARFT adopted AVS as a broadcast and TV standard.

[5]Of the roughly 175 AVS working group members, approximately 30 are foreign companies (Ernst, 2011).The AVS Working Group was organized in 2002 as a response to what were considered to be excessive royalty fees paid by Chinese DVD manufacturers using the MPEG-2 standard. Its formation was a result of initiatives taken by the Institute for Computer Science of the Chinese Academy of Sciences and the MII to bring together Chinese experts working on audio and video coding technologies to develop a standard suitable for Chinese conditions, especially the needs of Chinese industry.

make an *ex ante* commitment to license on FRAND or FRAND-RF terms, or participate in the AVS patent pool. Contributions to a standard of Chinese-held SEPs have only the latter two options available to them. Members are expected to participate in at least one subgroup, but may choose not to participate if they do not want to make a commitment to license a patent for the standard being developed by that subgroup. The policy also provides for a 90-day review period for assessing the relevance of non-contributed or disclosed patents (AVS, 2008). Left unaddressed in the original formulation were nonmember IPR, legacy patent issues, questions of compulsory licensing, and whether the "reasonableness" component of FRAND should have a defined upper limit (Huang and Reader, 2013).

By 2006, guidelines for the operation of an AVS patent pool were announced, including the establishment of a one RMB yuan per unit fee per device for licenses issued in China. The guidelines also provide for the establishment and responsibilities of a patent pool executive council which is composed of 5 representatives of government agencies, 6 members drawn from participating patent holders, 6 members of the Working Group who are users of the standard, the head of the Working Group, and the director of the patent pool management center. The patent pool operates on a not-for-profit basis. (Li, n.d.).

Today, AVS is supported by the Working Group, the patent pool management organization, and an AVS industrial alliance that seeks to promote the commercial deployment of the standard. As noted, AVS is seen by many observers as a progressive standard-setting organization that developed an IPR policy at an early stage of its development in an attempt to accommodate private interests and social benefits in an equitable fashion through open and transparent procedures and principles.

The AVS policy formed the basis of CESI's *IT Standard Drafting Organizations' IPR Policy Template*, first drafted in 2006. The template has now gone through 15 revisions involving a task force which has included representatives from China's AVS, RFID, and LINUX working groups, and which has also had the counsel of standards officers from leading multinational firms, including Intel, Microsoft and Sun Microsystems.[6] In an attempt to emulate international best practices, the template is intended to provide—and has provided—a framework for Chinese SSOs in the development of their IPR policies, while also recognizing that different SSOs may craft their policies in response to distinctive needs. The template provides a set of suggested definitions (e.g. "necessary claim") for key terms in the policies, an explication of the conditions for understanding IPR contributions to the standardization process, guidance on disclosure requirements, and suggested licensing options (FRAND-RF, patent pool, FRAND).

---

[6]CESI Standardization Development Research Center (n.d.), "IT Standard Drafting Organizations' IPR Policy Template."

CCSA has also attempted to provide policy guidance through its Intellectual Property Rights Policy, introduced in 2007 for trial implementation. It is a somewhat less specific document but also provides disclosure and licensing guidelines. Like the CESI template, it calls for FRAND-RF and FRAND options, but does not provide for patent pools. As a member of 3GPP, CCSA is somewhat constrained in developing IPR policies that deviate from those of 3GPP.

These initiatives at the CESI, CCSA and working group levels, should be seen in the context of a broader effort at the SAC level to formulate *national* policy guidelines. In 2004, SAC published a draft proposal for a patent policy that raised significant concerns among international stakeholders, and led SAC to reconsider its terms. In November 2009, in the face of a growing concern that there was little policy consistency in the SSO approaches to IPR matters, in spite of the CESI and CCSA initiatives, a revised "Proposed Regulations for the Administration of the Formulation and Revision of Patent Involving National Standards" was issued by SAC.

While noting improvements, comments from international stakeholders again called attention to provisions in the draft which seemed to be at odds with provisions found in the policies of international SSOs and which failed to provide adequate protection to the rights of IPR owners, especially with regard to licensing terms and to compulsory licensing in the case of mandatory standards (Willingmyre 2009, 2010). Shortly after the release of the "Proposed Regulations," in January, 2010, SAC's China National Institute of Standardization (CNIS) released what became known as the "Disposal Rules for the Inclusion of Patents in National Standards." Although the Disposal Rules were seen as removing some of the more troubling provisions of the Draft Regulations, some concerns remained such as a failure to distinguish between essential patents and essential patent claims, ambiguity over differences between patent declarations and actual licenses, and clarifications over disclosure obligations, especially with regard to nonparticipants (Willingmyre 2010). The draft policy again underwent a careful review and, as discussed below, a new draft "Regulatory Measures on National Standards Involving Patents (Interim)" was released on December 18, 2012.

As noted above, China's approaches to strategies for standardization and intellectual property development seek to serve Chinese interests in national technological development while striving for consistency with international norms and accommodating the interests of international stakeholders. Some Chinese initiatives have been confusing and troubling in light of international norms. Both the American Chamber of Commerce in China (AmCham) and the European Chamber have regularly expressed their concerns about the evolution of China's standardization and IP regimes.

A recent report from the European Chamber on patenting behavior in China also addresses a series of issues dealing with standard-setting procedures (Prud'homme, 2012). According to the report, in spite of greater opportunities for participation in Chinese standards organizations, foreign invested enterprises

(FIEs) are still denied access to some important technical committees. As a result, they "are unable to obtain information on the scope and requirements of patents to implement the standards that are frequently used in mandatory certification schemes." It goes on to note that (p. 11-12)

> European IP holders have continued to experience great difficulties in engaging the Chinese telecommunications industry in licensing discussions over "essential" patents, i.e. those containing one or more claims that are critical to the implementation of a technical specification or standard.

The report expresses a concern that Chinese approaches to standardization are being used to support indigenous technologies, often by using a standard that reflects the distinctive capabilities of Chinese enterprises. European companies report that they have difficulties in licensing discussions with Chinese counterparts over the determination of "essential" patents. Foreign stakeholders have also been concerned over the implementation of China's information security initiative, the Multi-Level Protection Scheme (MLPS), which "... includes domestic IP requirements that do not allow foreign companies to build a variety of Chinese infrastructure, whether as part of government procurement or commercial initiatives."[7]

AmCham has also expressed its concerns about Chinese standardization policies.[8] As with the European Chamber, AmCham has been concerned about Chinese initiatives for developing information security standards to the exclusion of international companies. It has also commented on Chinese reluctance to recognize as "international standards" those standards developed by U.S.-based firms which have not been approved by ISO, ITU, and IEC in spite of the fact that they are often otherwise globally accepted and meet WTO requirements for international standards. Like the European Chamber, AmCham has also expressed concern about rights of participation in some Chinese technical committees in spite of SAC regulations intended to allow foreign invested enterprises registered in China to participate and vote in technical committees.

AmCham also reports that its member companies are often concerned that Chinese standard-setting procedures do not adequately protect proprietary information, including concerns that the copyrights and patents of U.S. SSOs are not adequately protected against infringement.[9] AmCham has urged SAC to cooperate more fully with international SSOs in order to bring its IPR policies up to the best international practices, and to work more closely with the State

---

[7] See also the case studies of MLPS and related policies in Ahrens (2012), Ernst (2011), and Ernst and Martin (2010).

[8] See various white papers at AmCham China, available at http://www.amchamchina.org/whitepaper.

[9] Thus, while foreign companies express a desire to participate more fully in Chinese standards bodies, they are also concerned that participation carries IPR risks.

Intellectual Property Office (SIPO) and the National Copyright Administration of China (NCAC) (AmCham, 2012, 2013).

Questions about standard-essential patents in China are also shaped by the broader Chinese legal context. This includes provisions of the Patent Law requiring compulsory licensing when the "public interest" (largely unspecified) is at stake and in cases where the patent owner fails to "sufficiently exploit" (again largely unspecified) the patent. China's Antimonopoly Law has also been a matter of concern to foreign companies who fear that they could be held in violation of that law for failing to license technology that may be needed for the innovation needs of a third-party, potentially establishing market dominance under Chinese definitions (AmCham, 2012).

It is not entirely clear, as of this writing, how fully the newly released SAC "Regulatory Measures on National Standards Involving Patents (Interim)" will address these concerns.[10] The "Measures" document is quite brief and states principles in rather general terms. Its short provisions on disclosure require participants in working groups to declare essential patents, although the provision is vague as to timing and whether full patent searches are expected. Participants are given three licensing options: choose FRAND, FRAND-RF, or state an unwillingness to license according to the first two terms. If the last is chosen, the standard cannot be based on the patent(s) in question. This is consistent with similar provisions in many western SSO IPR policies. The state reserves the right to suspend the use of the standard in cases where an undisclosed patent has been included in the standard. The provisions called for a transfer of obligations in the event of a change in the ownership of the patent. More specifically, where a patent holder that has made a licensing declaration transfers the declared patent, it must have the transferee agree to be bound by that declaration. Again, this is consistent with similar provisions in ETSI, ITU and other SSO IPR policies.

On the question of mandatory standards, the "Measures" skirt around the question of compulsory licensing by noting that mandatory standards and principles should not incorporate patented technology. When they do, however, the policy calls for negotiations between the holder of the patent and the state, with a suspension of the publication of the standard until the issue has been resolved. In cases where a mandatory standard does involve patents, the state is expected to publish a text of the standard and information about the IPR for 30 days, during which time stakeholders are encouraged to submit additional information.

The draft "Regulatory Measures" were issued with an invitation for comments by stakeholders. As of this writing, a number of foreign groups have submitted views on the draft, including The U.S. China Business Council, the American Intellectual Property Law Association, the American Bar Association, and the Intellectual Property Owners Association. While most stakeholder groups expressed support for the SAC initiative, they also submitted a number

---

[10]An English translation is available at http://sunsteinlaw.com/wp/wp-content/uploads/2013/01/2013_01_IP_Update_PRC.pdf.

of suggestions concerning definitional clarity, disclosure requirements, licensing requirements, and additional procedural transparency.

In addition to the new SAC "Regulatory Measures," there have also been recent reports that CESA and the Electronics Industry Association are anxious to resolve uncertainties about policies for standard essential patents, and are working with the Intellectual Property Center of MIIT on the development of a training program using policy principles developed by ETSI.

China's evolving competition policy is also shaping the ways in which SEP issues are being approached. For instance, the State Administration for Industry and Commerce (SAIC) has recently issued a sixth draft for comments on its "Provision of the Administrative Authorities for Industry and Commerce on Prohibition of Abuse of Intellectual Property to Exclude or Restrict Competition," a document intended to provide guidance for the administration of China's Antimonopoly Law, and which includes reference to standard-setting practices.

In addition, in an important case decided by the Shenzhen Intermediate People's Court on February 4, 2013, Huawei was awarded $US 3.2 million in damages in two suits against InterDigital. These suits alleged that the latter had used its dominant market position to deny licenses to Huawei on FRAND terms. In the first case, Huawei alleged that InterDigital had a dominant market position in China and the United States in the market for the licensing of essential patents it owned and abused its market power by engaging in differentiated pricing, tying, and refusal to deal.

In connection with this suit, the Shenzhen court held that InterDigital violated China's Antimonopoly Law by (1) making proposals for royalties from Huawei that the court believed were excessive; (2) tying the licensing of essential patents to the licensing of non-essential patents; (3) requesting as part of its licensing proposals that Huawei provide a grant-back of certain patent rights to InterDigital; and (4) commencing a United States International Trade Commission (USITC) action against Huawei while still in discussions with Huawei for a license (InterDigital Annual Report, 2012; Nylen and Swift, 2013).

In the second suit, Huawei argued that the FRAND commitments InterDigital made to ETSI obligated it under Chinese law to negotiate with Huawei on FRAND terms, and requested the court to determine a FRAND rate for licensing SEPs to Huawei. The court ruled that royalties for InterDigital patents essential to implementing 2G, 3G, and 4G standards should not exceed 0.019 percent of the price of individual Huawei products, but has not yet made public the reasoning for deciding on this rate (InterDigital Annual Report, 2012).

In both the Shenzhen court decision and the SAIC draft, we see how competition policy may introduce a policy bias against the interests of rights holders, a bias which would be troubling to some foreign corporations. But they also point to a still evolving situation, calling for the further clarification of the IPR policies of Chinese SSOs.

Policy and institutional uncertainty

Over the past decade, Chinese approaches to standardization, including policies for essential patents in standards, have continued to evolve in ways that show sensitivity to international norms and the interests of international stakeholders, including multinational companies in some circumstances, although in some instances have excluded foreign company participation. At the same time, the commitment to building a standards regime to serve Chinese national interests remains powerful. Thus, the development of policies for standard-essential patents is still a work in progress and continues to reflect the problems of accommodating the competing objectives noted at the outset.

These problems are made more complex by ongoing institutional uncertainty. The recent 18th Congress of the Chinese Communist Party has brought into office a new group of leaders who express optimism over China's future but also recognize the daunting challenges they face. The new party chairman, Xi Jingping, has celebrated China's recent progress and referred to China as being in a period of "national revival" or "rejuvenation" (*fu xing*). National technological development to enhance national capabilities for innovation is central to this revitalization in the minds of Chinese leaders. Yet in spite of aggressive policies to promote innovation, including rapidly increasing R&D expenditures, there is a widespread disappointment in actual achievements, and a recognition that existing institutions are not serving Chinese aspirations, witness the weak record of producing successful standards and high-quality exploitable patents. In recent months there have been growing discussions of further institutional reform. While detailed reform proposals have yet to appear, it is likely that a number of the institutions involved with standardization and intellectual property policies will be affected.

In addition to this institutional uncertainty, approaches to standard-essential patents may also be affected by changing attitudes towards intellectual property. As suggested above, Chinese interests in ICT standardization have been strongly influenced by concerns over licensing fees and the overall economic benefits redounding to those controlling standards and owning the intellectual property. In this sense, China has sought to secure what it considers to be a more equitable access to standards in order to better serve its manufacturing activities; monetizing IP and standards has not been the primary goal (Breznitz and Murphree, 2012). As China's patent portfolio has expanded, however, so too has its patent litigation and there are signs that it may be moving towards a much greater concern for IP monetization. If so, we may see the further growth of new markets for IP and the emergence of new players such as non-practicing entities (NPEs). These developments would introduce significant new dynamics to issues of standards and IPR policy in China.

8.3 India

Like China, India has been concerned about the benefits flowing to Western IP owners and the questions of equity concerning their distribution. Also like China, its standards regime has had to make adjustments in response to WTO imperatives. Unlike China, however, India had, until recently, shown less of a strategic orientation towards developing its own standards and intellectual property, preferring instead to rely on established international standards, and on international practice for developing policies affecting patents in standards. More recently, though, awareness that its huge market and distinctive social and physical conditions present interesting technological opportunities has prompted much more attention to Indian standards development and the creation of Indian intellectual property. Indian thinking about a more strategic approach to standards development has also been influenced by concerns that imports from China are capturing market share from Indian firms. Further, India's firms wish to raise quality and save testing and certification costs by complying with trusted standards, while public authorities tend to focus support policies on higher-quality enterprises.

This is evident, for instance, in the Government of India National Telecom Policy, introduced in 2012, which focuses not only on a significant expansion of telecom services, but also calls attention to the need to stimulate innovation in Indian industry and develop new strategies for promoting Indian telecom technology. In language reminiscent of Chinese policy discourse, the new Policy calls for numerical targets for the growth and self-sufficiency of the industry, and the creation of "... a roadmap to align technology, demand, standards and regulations for enhancing competitiveness of domestic manufacturing" (Government of India, 2012). It calls attention to the importance of procurement preferences for "...domestically manufactured telecommunication products...," and the need to promote indigenous R&D and IPR creation as part of the current 12th five-year plan period. As part of this effort, India should:

> Develop and establish standards to meet national requirements, generate IPRs, and participate in international standardization bodies to contribute in (the) formulation of global standards, thereby making India a leading nation in the area of international telecom standardization. This will be supported by establishing appropriate linkages with industry, R&D institutions, academia, telecom service providers and users. (Government of India, 2012)

Key institutions

The Bureau of Indian Standards (BIS) is the Indian government's main institution for standardization. Established in 1987, superseding the Indian Standards Institution established in 1947, BIS operates 14 industry related sectors, each of which is managed by a "division council." Of greatest relevance to this study is the Electronics and Information Technology Division Council (LITD), created in 1977, which includes 21 "sectional committees" covering various fields of IT, including computer communications, networks and interfaces. As with other BIS division councils, LITD committees have attempted to harmonize their standards with those in ITU, IEC, and ISO.

BIS reportedly has not yet developed its own IPR policies (Ramakrishna, et. al, 2012). Since many of its standards are technically equivalent to international standards, the BIS position has been to rely on the IPR policies of the international standards organizations that developed the standard. BIS leaves it to manufacturers wishing to use a standard to negotiate license terms if IPR is an issue.

A second important government body is the Telecommunication Engineering Center (TEC) of the Department of Telecommunications of the Indian Ministry of Communications and Information Technology. TEC has a lead role in the development of standards for telecom equipment, services, and interoperability. It maintains regular interaction with ETSI and the ITU and participates in a number of other international standards setting bodies, such as ITU-T, the WiMAX forum, IETF, IEEE, and the like. As with standards bodies under BIS, it appears that TEC also does not have a well-established IPR policy. A good bit of its work is to develop specifications for equipment to be used under Indian conditions, with most of this equipment conforming to ITU standards. Thus, TEC looks to the IPR policy of the ITU in developing its standards and specifications. TEC is also supporting the development of a new national telecom standards development organization, called for in the National Telecom Policy. In May 2013, the establishment of a new Telecom Standards Development Society of India (TSDSI) was announced with membership drawn from diverse stakeholders including manufacturers, service providers, research and academic institutions and government organizations (Department of Telecommunications, 2013).

In addition to these two mainline governmental institutions, two non-governmental standards bodies should be noted. The Global ICT Standardization Forum for India (GISFI), founded in 2008, seeks to provide greater coherence to ICT standardization in India in such emerging fields as energy, telemedicine, wireless robotics, and biotechnology, and to integrate more fully Indian ICT standards initiatives with international trends. Its members include Indian and foreign firms and Indian research institutions, but it seeks participation from the full array of stakeholders.[11] It maintains working groups in the areas of information security and privacy, future radio networks, the Internet, cloud and service-oriented networks, green ICT, and spectrum. In December 2011, GISFI

---

[11]Corporate members include Niksun, NEC, Ericsson, Tejas Networks, Motorola, Tata Consultancy Services Limited, Huawei, Nokia Siemens Networks, Veriserve, and Samsung. For a complete list, see http://www.gisfi.org/membership.php.

IPR and Standards in Emerging Economies

cooperated with the ITU in sponsoring a workshop on standards and intellectual property rights.

GISFI maintains an IPR policy based on that of ETSI, but having its own features as well. The policy seeks to reduce economic and legal risks to stakeholders and to balance the interests of rights holders and the needs of the public. It calls for timely disclosure but does not obligate members to engage in patent searches. When SEPs are brought to the attention of GISFI, its Director General is expected to request from the patent owner an irrevocable commitment to grant licenses on FRAND terms. The owner is also expected to make reasonable efforts to notify a new assignee of GISFI standards commitments in the event of a transfer.

In the event that the patent owner refuses to license, GISFI will explore the availability of alternative technologies for the standard. If an alternative technology does not exist, and if the patent owner is a member of GISFI, the Director General will ask for a reconsideration of the licensing decision, and if the decision is upheld, a written explanation of the owner's decision not to license within three months of the request. The explanation, and supporting information, will then be sent to the GISFI General Assembly for consideration. If the patent owner is not a member, the Director General shall contact the patent owner requesting an explanation for the licensing decision and requesting reconsideration. If the decision not to license is upheld, the matter will be referred to the General Assembly for reconsideration, with counsel, of whether the patent in question is essential for the standard. In cases where a license is not available after the publication of the standard, the Director General will again take the initiative to contact the patent owner with requests for an explanation and reconsideration. Thereafter, the matter will again be referred to the General Assembly for review, with members urged to use their good offices to find a solution.

> A second non-governmental organization of interest is the Development Organization of Standards for Telecommunications in India (DOSTI), a private SDO committed to the development of telecom standards suitable for Indian conditions. It currently has eight working groups and a membership that includes both Indian entities and foreign companies.[12]

DOSTI maintains an IPR bearing a resemblance to that of GISFI. It also calls for timely disclosure but carries no obligation for full patent searches. Owners of SEPs will be requested by the DOSTI Director General to grant irrevocable licenses on FRAND terms. In the event of transfers, the patent owner is expected to notify the assignee of any commitments made to DOSTI. The

---

[12]Members currently include Cellular Operators Association of India (COAI), Association of Unified Telecom Service Providers of India (AUSPI), Tejas Networks, TCOE India, CEWiT, Ericsson, Qualcomm, Nokia Siemens Networks, IIT Bombay, IIM Ahmedabad.

DOSTI policy provides for cases where transferability is disputed, including the conduct of an IPR policy search if requested by the Department of Telecommunications.

In the event of non-availability of licenses, the DOSTI Executive Council will consider whether an alternative technology is available. If not, work on the standard will be discontinued and the Director General will ask the members to reconsider. If the decision not to license is maintained, the member will be asked to submit in writing the reasons for refusing to license within three months, after which the matter will be referred to the Executive Committee.

In cases where the IPR owner is not a DOSTI member, the policy calls for efforts to be made through direct DOSTI contacts and through the good offices of members to have the owner reconsider or present in writing the reasons for refusing to license. If there is no reconsideration or if the owner refuses to respond within three months, the matter is referred to the Executive Committee for its consideration. This may lead to the referral of the matter to the appropriate working group to consider a technological workaround or to the Department of Telecommunications for appropriate action, including perhaps the non-recognition of the standard. The DOSTI policy also provides for IPR owned by DOSTI itself, including FRAND licensing provisions (Ramakrishna, et. al, 2012).

Most recently, the government of India announced the development of the Telecom Standards Development Society, India (TSDSI).[13] TSDSI will bring together stakeholders from industry, academia, and government organizations in "…an autonomous body which will drive consensus regarding standards to meet national requirements."

e-Government and information security

Due to its diverse and often incompatible legacy government information systems, the government of India faces significant challenges in providing for effective e-government services. In its efforts to modernize e-government, it has introduced an open standards policy which includes a mandatory royalty-free approach to licensing, and requirements that the standard have a technology neutral specification and be adapted and maintained by a not-for-profit organization. In the event that an open standard meeting these conditions is not available, the policy also provides for interim standards, the IPR of which can be licensed on a FRAND basis, and which relaxes the not-for-profit organization requirement (Government of India, 2010). The e-government policy also includes guidelines for information security. It establishes an e-Governance Security Assurance Framework based on ISO 27001 consistent with the U.S. Information Security Program for Federal Information Systems (Ramakrishna, et. al, 2012).

---

[13] http://www.tta.or.kr/include/Download.jsp?filename=externalDocument/GSC17-PLEN-84_India_s_Statement_at_GSC_Korea.docx.

India in transition

The growing importance of India in the global ICT industries, based on its market size and the growing technological capabilities of the Indian technical community, is, as in China, leading to new attention to standards and intellectual property rights. This has affected not only the behavior of Indian stakeholders, but has also led to the growth of standard-related activities involving international standards bodies and corporations as well. IEEE, for instance, has established a "standards interest group" (SIG) for India with the goal of stimulating greater Indian involvement in the IEEE global standards process (IEEE, 2011). GISFI has been active in working with ITU. Interestingly, the Chinese firm Huawei, which has an active presence in India and is a GISFI member, recently sponsored a joint ITU-GISFI workshop on "Bridging the Standardization Gap."

It is notable that explicit attention to the development of an IPR policy for standardization is found more in India's nongovernmental SSOs than in the government standards agencies. Indian approaches to the development of policies for IPRs in standards have tended to follow those of established international standards bodies, and this is likely to continue. As indicated in the National Telecom Policy, however, there seems to be a growing appetite for government supported indigenous Indian standards development efforts incorporating Indian intellectual property. India is thus at an interesting transitional point where the reconciliation of emerging Indian industrial policy objectives and harmonization with international practices are likely to face new challenges.

8.4 Brazil

Like India and China, Brazil is an important emerging economy committed to moving up the value chain through greater attention to enhancing technological capabilities.[14] Also like India, however, it has shown less of a strategic orientation towards standards and IPR development than China and its standards regime is only gradually coming to grips with issues such as defining and licensing SEPs. It appears, as of this writing, that Brazil does not have any well-developed policies for dealing with IPR in standards (Barbosa, 2012).

Key institutions

Brazil's National System of Metrology, Standardization and Industrial Quality (SINMETRO) and its patent and trademark office (INPI), are both part of the Ministry of Development, Industry and Foreign Trade. The SINMETRO system includes the National Council of Metrology, Standardization and Industrial Quality (CONMETRO), the National Institute of Metrology, Standardization, and Industrial Quality (INMETRO) and the Brazilian Association of Technical Norms

---

[14] A key development was the passage of the 2004 Brazilian Innovation Law.

(ABNT), which serves as Brazil's main organization for standardization. ABNT is a private organization but since 1940 has been officially recognized as the national standards body and receives public as well as private funding. It develops standards through its own technical committees and also accredits sectoral standardization bodies. SINMETRO and ABNT policies call for the use of international standards to the extent possible and as references for local work on standardization and technical specifications.

Although Brazil has yet to develop explicit policies for IPR in standards, the relationships between standard-setting and intellectual property are becoming more prominent. Brazil's 2010 public procurement law, for instance, allows for the consideration of technology and industrial policy objectives in procurement decisions, including purchasing decisions in support of the development of national standards in ICT and the promotion of open standards for e-government.[15]

The Brazilian patent office, INPI, plays a role in IPR in standards issues. It is empowered to analyze and approve license payments, including those related to standards. Under Brazilian law, patents can only be transferred to new assignees following specific action by the Patent and Trademark Office. Competition policy and the role of the Competition Administrative Court (CADE) are also significant in how IPR in standards is treated, especially with regard to patent pools and royalty rates (Barbosa, 2012).

8.5 Conclusions and Recommendations

In a global economy where intellectual property has become an increasingly valuable asset, it is not surprising that key emerging economies are taking the creation, protection, and utilization of that asset to be matters of national and corporate importance. Likewise, the growing strategic importance of technical standards is also attracting national and firm-level policy attention. We should not be surprised, therefore, that the role of intellectual property in standard-setting will assume a more central place in the evolution of the policy and legal systems of these three emerging countries as it clearly already has in China. Given the growing size and importance of these economies, the ways in which they approach the development of their domestic IPR and standards regimes will have important implications for the norms and institutions by which international standardization and IPR affairs are governed.

In each of the countries reviewed, we see interesting tensions between tendencies toward harmonization with international norms and practices and tendencies towards more techno-nationalist agendas supporting the protection of national industries and the building of national champions. This tension is perhaps most evident in China, where a sense of the strategic importance of standards and IPR is more developed, but it is certainly evident in the evolving industrial policies of India and Brazil as well.

---

[15] Brazil, like China, has not joined the WTO Government Procurement Agreement.

IPR and Standards in Emerging Economies 139

In short, these three emerging economies are all working towards national regimes for managing the relationships between IPR and standards in ways that serve national interests but which also accord with international best practices. The integration of these objectives, in the first instance, will be affected by the evolution of government-industry relations, especially with regard to the role of the state, as opposed to industrial enterprises, in taking the lead in standard-setting activities. This question, in turn, is a function of a deeper set of issues involving political traditions and state-society relations which are beyond the scope of this report. But, in addition, the evolution of these national regimes will also be strongly influenced by the development of the intellectual property assets which stakeholders in these countries bring to domestic standard-setting and international standardization activities. The weak patent portfolios that, until recently, have been characteristic of most firms in these countries have ensured that they have in effect been "standards takers" rather than "standards makers." As technological development proceeds and domestic patent portfolios expand we can expect that this situation will change, and with it, more robust approaches to standardization will emerge along with more focused attention to the treatment of standard-essential IPR.

In all three countries reviewed, the development of a modern technical standards regime is still a work in progress. This is true even for China, where its learning curve is notably steep and it has shown a far more robust approach to building a national standardization system than has India or Brazil. While there are limits to how much the U.S. government can contribute to the development of these standards regimes, the fact that they are all in varying stages of formation suggests that there are possibilities for mutually beneficial interactions, especially with regard to education, training and raising awareness as to the importance of developing IPR policies in the early stages of building SSO capabilities. As the AVS case illustrates, an awareness of international practices combined with good counsel from knowledgeable foreign experts can lead to the development of positive policies and procedures which avoid pitfalls experienced by others.

Recommendation 8:1

The U.S. government, should explore ways to promote awareness of the importance of developing IPR policies at an early stage of the development of SSOs in these and other emerging economies, and should, in conjunction with non-governmental standards entities, explore ways to offer training programs for those working to develop their organizations and the policies needed for successful national standardization.

Recommendation 8:2

In the meantime, agencies of the United States government, such as the United States Patent and Trademark Office, the Office of the United States Trade

Representative, and the National Institute of Standards and Technology should closely monitor and report on continuing developments in these countries and other major emerging economies regarding standard-setting and the management of intellectual property.

References

Adams, J., D. Pendlebury and B. Stembridge. 2013. Building Bricks: Exploring the Global Research and Innovation Impact of Brazil, Russia, India, China and South Korea. Thomson Reuters. Available: http://sciencewatch.com/sites/sw/files/sw-article/media/grr-brick.pdf.

Ahnrens, N. 2012. National Security and China's Information Security Standards. Washington D.C.: Center for Strategic and International Studies.

Almunia, J. 2012. Competition Policy for Innovation and Growth: Keeping Markets Open and Efficient. Available: http://europa.eu/rapid/press-release_SPEECH-12-172_en.htm?locale=en.

AmCham China. 2010, 2011, 2012, 2013. White Papers Available: http://www.amchamchina.org/whitepaper.

American Bar Association. 2007. Standards Development Patent Policy Manual. Chicago: American Bar Association.

American National Standards Institute. Rev 2008. 3.1 ANSI Patent Policy - Inclusion of Patents in American National Standards. Washington, D.C.: ANSI.

An, B. 2012. The Global Governance of Standardization: The Challenges of Convergence. Research Center for Chinese Politics and Business Working Paper No. 32. Bloomington, IN: Indiana University.

AVS. 2008. Intellectual Property Rights Policy of the Audio Video Coding Standard Working Group of China. Available: http://www.avs.org.cn/en.

Barbosa, D. 2012. Intellectual Property and Standards in Brazil. Available: http://sites.nationalacademies.org/PGA/step/PGA_058712.

Bekkers, R. and A. Martinelli. 2012. Knowledge positions in high-tech markets: trajectories, standards, strategies and true innovators. Technological Forecasting & Social Change. 79:1192-1216.

Bekkers, R. and A. Updegrove. 2012. A Study of IPR Policies and Practices of a Representative Group of Standards-Setting Organizations Worldwide. Available: http://www.nap.edu/catalog.php?record_id=18510.

Blind, K. 2011. Study on the Interplay between Standards and Intellectual Property Rights (IPRs). Tender No: ENTR/09/015 OJEU S136 of 18/07/2009, Final Report. Available: http://ec.europa.eu/enterprise/policies/european-standards/files/standards_policy/ipr-workshop/ipr_study_final_report_en.pdf.

Breznitz, D. and M. Murphree. 2012. Shaking Grounds? Technology Standards in China. Available: http://sites.nationalacademies.org/PGA/step/PGA_058712.

Chia, T. 2012. Fighting the smartphone patent war with RAND-encumbered patents, Berkeley Tech. L.J. 27: 209-240.

Clougherty, J. and M. Grajek. 2012. International Standards and International Trade: Evidence from ISO 9000 Diffusion. National Bureau Economics Research Working Paper 18132. Cambridge, MA.

Contreras, J. 2013. Technical standards and ex ante disclosure: results and analysis of an empirical study. Jurimetrics J. 53:163-211.
Contreras, J. 2013. Fixing FRAND: a pseudo-pool approach to standards-based patent licensing. Antitrust L.J. 79:1-43.
Contreras, J. 2012. Survey of Bioinformatics Standards. Available: http://sites.national academies.org/PGA/step/PGA_058712.
Contreras, J. and C. McManis. 2012. Materials Sustainability Standards and Intellectual Property. Available: http://sites.nationalacademies.org/PGA/step/PGA_058712.
Damien Geradin, Anne Layne-Farrar, and A. Jorge Padilla, The Complements Problem Within Standard Setting: Assessing The Evidence On Royalty Stacking, *B.U. J. Sci. & Tech. L.*, Vol. 14:144-176.
DeNardis, L. 2012. E-government Acquisition Processes in the U.S., EU, India and Japan. Available: http://sites.nationalacademies.org/PGA/step/PGA_058712.
Department of Telecommunications. 2013. Statement of Intent by India, Global Standards Collaboration Conference, Jeju Island, Korea. Available: http://www.tta.or.kr/include/Download.jsp?filename=externalDocument/GSC17-PLEN84_India_s_Statement_at_GSC_Korea.docx.
Ernst, D. 2006. Innovation Offshoring: Asia's Emerging Role in Global Innovation Networks. East-West Center Special Reports Number 10. Honolulu, HI.
Ernst, D. 2011. Indigenous Innovation and Globalization: The Challenge for China's Standardization Strategy. La Jolla, CA: University of California Institute of Global Conflict and Cooperation. Honolulu, HI.
Ernst, D. and S. Martin. 2010. The Common Criteria for Information Technology Security Evaluation – Implications for China's Policy on Information Security Standards. East-West Center Working Paper No. 108. Honolulu, HI.
European Telecommunications Standards Institute Rules of Procedure. Annex 6: ETSI Intellectual Property Rights Policy. 2013. Available: http://www.etsi.org/images/files/IPR/etsi-ipr-policy.pdf.
European Commission. Directorate-General for Competition Press Release. 2013. Antitrust: Commission sends Statement of Objections to Motorola Mobility on potential misuse of mobile phone standard essential patents. Brussels. Available: http://europa.eu/rapid/press-release_IP-13-406_en.htm.
European Commission, Directorate-General for Competition Press Release. 2012. Commission Sends Statement of Objections to Samsung on potential misuse of mobile phone standard-essential patents. Brussels. Available: http://europa.eu/rapid/press-release_IP-12-1448_en.htm.
European Commission, Directorate-General for Competition Memo. 2012. Samsung – Enforcement of ETSI Standards Essential Patents (SEPs). Brussels. Available: http://europa.eu/rapid/press-release_MEMO-12-1021_en.htm.
European Commission, Directorate-General for Competition Press Release. 2012. Commission Opens Proceedings Against Motorola. Brussels. Available: http://europa.eu/rapid/press-release_IP-12-345_en.htm.
European Commission. 2012. Google-Motorola Mobility Merger Procedure Article 6(1)(b) Decision. Case No. Comp/M. 6381. Luxembourg. Available: http://ec.europa.eu/competition/mergers/cases/decisions/m6381_20120213_20310_2277480_EN.pdf.
European Commission. O.J. 2011. Communication from the Commission on Guidelines on the Applicability of Article 101 of the Treaty on the Functioning of the EU to Horizontal Co-operation Agreements. Available: http://eur-lex.europa.eu/LexUriServ/LexUriServ.do?uri=OJ:C:2011:011:0001:0072:EN:PDF.

European Commission. 2010. Guidelines on the applicability of Article 101 of the Treaty on the Functioning of the European Union to horizontal co-operation agreements, SEC 528/2. Brussels. Available: http://ec.europa.eu/competition/consultations/2010_horizontals/guidelines_en.pdf.

Executive Office of the President. 2013. Patent Assertion and U.S. Innovation. The White House, Washington, D.C. Available: http://www.whitehouse.gov/sites/default/files/docs/patent_report.pdf.

Farrell, J. J. Hayes, C. Shapiro, and T. Sullivan. 2007. "Standard setting, patents, and hold-up. Antitrust Law Journal. 74:603-670.

Gallini, N. and B. D. Wright. 1990. Technology transfer under asymmetric information. RAND Journal of Economics. 21(1):147-160.

Gilbert, R. 2011. Deal or no deal? licensing negotiations in standard-setting organizations. Antitrust L. J. 77:855-888.

Global ICT Standardisation Forum for India. GISFI Membership List. Available: http://www.gisfi.org/membership.php.

Goudelis, M. 2012. Presentation on EPO Cooperation with Standards Development Organisations. Available: http://sites.nationalacademies.org/PGA/step/PGA_072825.

Government of India. Ministry of Communication and Information Technology. Department of Information Technology. 2010. Policy on Open Standards for e-Governance. Available: http://egovernance.gov.in/policy/policy-on-open-standards-for-e-governance/Policy%20on%20Open%20Standards%20for%20e-Governance.pdf.

Government of India. Ministry of Communication and Information Technology. Department of Information Technology. 2012. National Telecom Policy 2012. Available: http://www.dot.gov.in/ntp/NTP-06.06.2012-final.pdf.

Hesse, Renata. 2012. Six "Small" Proposals for SSOs Before Lunch (Remarks as Prepared for the ITU-T Patent Roundtable). United States Department of Justice. Washington, D.C. Available: http://www.justice.gov/atr/public/speeches/287855.pdf.

Huang, T. and C. Reader. 2013. China's ABS Intellectual Property Rights Policy - A New Approach for Developing Open Standards. Available: http://www.google.com/url?sa=t&rct=j&q=&esrc=s&source=web&cd=1&sqi=2&ved=0CDEQFjAA&url=http%3A%2F%2Fcdn.nbr.org%2Fdownloads%2FCS09_HUANG_slides_EN.ppt&ei=GN_UUOfiMMjxigL6lYD4DQ&usg=AFQjCNGq3lItnZaPOmBgV5Mxtyl2qMIqkw&bvm=bv.1355534169,d.cGE&cad=rja.

Institute of Electrical and Electronics Engineers-Standards Association. 2011. IEEE Announces Standards Interest Group (SIG) for India: Move to Propel India's Involvement in the IEEE Global Standards Process. Available: http://standards.ieee.org/news/2011/sig.html.

International Telecommunications Union. 2012. Guidelines for Implementation of the Common Patent Policy for ITU-T/ITU-R/ISO/IEC 23/04/02. Available: http://www.itu.int/dms_pub/itu-t/oth/04/04/T04040000010003PDFE.pdf.

International Telecommunications Union. 2012. Joint ITU-GISFI Workshop on Bridging the Standardization Gap: Sustainable Rural Communications. Available: http://www.itu.int/en/ITU-T/Workshops-and-Seminars/bsg/Pages/default.aspx.

InterDigital, Inc. Invention Collaboration Contribution: InterDigital Annual Report. 2012. Delaware. Available: http://files.shareholder.com/downloads/IDCC/2696817853x0x657871/3375CC54-04CA-4694-83E4-DEBCB7508F97/2012_Annual_Report_and_2013_Proxy_Statement.pdf.

Ipeg. 2011 *Nokia v. IPcom* Ongoing UMTS Patent Litigation in Germany. Available: http://www.ipeg.eu/nokia-vs-ipcom-ongoing-umts-patent-litigation-in-germany/.

Jillavenkatesa, A., H. Evans, and H. Wixon. 2012. Patents and Intellectual Property Management in Nanotechnology: A NIST Perspective. Available: http://sites.national academies.org/PGA/step/PGA_072825.

Kesan, Jay and C. Hayes. 2012. Patent Transfers in the Information Age: FRAND Commitments and Transparency. Available: http://sites.nationalacademies.org/PGA/step/PGA_058712.

Kai-Uwe K., F. Scott Morton, and H. Shelanski. 2013. Standard Setting Organizations Can Help Solve the Standard Essential Patents Licensing Problem. Competition Policy International Antitrust Chronicle. Available: https://www.competitionpoli cyinternational.com/assets/Free/ScottMortonetalMar-13Special.pdf.

Kramler, T. 2012. FRAND Commitments and EU Competition Law. Available: www.itu.int/dms_pub/itu-t/oth/06/5B/T065B0000360016PPTE.ppt.

Layne-Farrar, A., J. Padilla, and R. Schmalensee. 2007. Pricing Patents for Licensing in Standard-Setting Organizations: Making Sense of FRAND Commitments. CEMFI Working Paper No. 0702. Madrid. Available: ftp://ftp.cemfi.es/wp/07/0702.pdf.

Lemley, M. and C. Shapiro. 2007. Patent holdup and royalty stacking. Texas Law Review 85:1991-2049.

Lemley, M. 2002. Intellectual Property Rights and Standard-Setting Organizations. Calif. L. Rev. 90:1889-1980.

Li, W. (n.d.). The Intellectual Property Issue in the Standardization Process: A Case Study From China's Experience. Available: http://www.thebolingroup.com/unifi er_divider/powerpoint_slides/wenwenli.pdf.

Majoras, D. 2012. Dissenting Statement of Chairman Majoras In the Matter of Negotiated Data Solutions, LLC, File No. 0510094. Federal Trade Commission, Washington: D.C.

Majoras, D. 2005. Recognizing the Procompetitive Potential of Royalty Discussions in Standard Setting. Federal Trade Commission, Washington, D.C.

Maskus, K. 2012. Private Rights and Public Problems: The Global Economics of Intellectual Property in the 21st Century, Washington, D.C.: Peterson Institute for International Economics.

Merrill, T. and H. E. Smith. 2000. Optimal Standardization in the Law of Property: The Numerus Clausus Principle, Yale L.J. 110:1-70.

Mock, V. 2012. Samsung Drops Injunctions Applications Against Apple. *The Wall Street Journal*. Available: http://online.wsj.com/article/SB10001424127887324407504578 187043081010804.html.

Moenius, J. 2004. Information versus Product Adaptation: The Role of Standards in Trade. Working Paper. Northwestern University Kellogg School, Evanston IL.

Nylen, L. and M. Swift. 2013. Chinese Court Eye Proper Royalty Rates for Patents Essential to Industry Standards. Available: http://www.mlex.com/US/content.aspx?ID=360997.

Office of Management and Budget. 1998. OMB Circular A-119; Federal Participation in the Development and Use of Voluntary Consensus Standards and in Conformity Assessment Activities. Final Revision. 63 FR 8546. Available: http://www.white house.gov/omb/circulars_a119.

Ono, N. 2012. Avoid Japanization: Lessons from Japanese Gridlock on Patent Assignment Recordation System. Available: http://dx.doi.org/10.2139/ssrn.2015119.

Organization for Economic Cooperation and Development. 2008. The Internationalisation of Business R&D: Evidence, Impacts and Implications. Paris: OECD.

Organization for Economic Cooperation and Development. 2012. OECD Science, Industry, and Technology Outlook. Paris: OECD.

References

Prud'homme, D. 2012. Dulling the Cutting-Edge: How Patent-Related Policies and Practices Hamper Innovation in China. European Chamber. Available: http://www.europeanchamber.com.cn/documents/confirm/en/52175d3bdf80f/en/pdf/14.

Raes, S. 2010. Who needs standards related patent registers and what should they look like? Presentation at the European Commission Workshop on Tensions between IPR & Standards. Brussels.

Ramakrishna, T., S.K. Murthy, and S. Makhotra. 2012. Intellectual Property in ICT Standards in India Available: http://sites.nationalacademies.org/PGA/step/PGA_058712.

Ramirez, E. 2012. Prepared Statement of the Federal Trade Commission before the United States Committee on the Judiciary Concerning Oversight of the Impact on Competition Orders to Enforce Standard-Essential Patents. Federal Trade Commission, Washington, D.C. Available: http://www.ftc.gov/speeches/ramirez/120711sep-stmt offtc.pdf.

Rizzolo, M. 2013. Essential Patent Blog. Huawei, ZTE Seek Stay of InterDigital 3G/4G ITC Investigation. Available: http://essentialpatentblog.com/2013/02/huawei-zte-seek-stay-of-interdigital-3g4g-itc-investigation/.

Robinson, F. and A. Torello. 2011. *Wall Street Journal*, 2nd Update: EU: Requested Apple, Samsung Mobile Patent Data.

Robinson, G. 2004. Personal Property Servitudes. U.Chi. L.Rev. 71(4):1449-1523.

Salant, D.J. 2009. Formulas for Fair, Reasonable, and Non-Discriminatory Royalty Determination. International Journal of IT Standards and Standardization Research. 7(1): 66-75.

Scott-Morton, F. 2012. The Role of Standards in the Current Patent Wars. U.S. Department of Justice, Washington: D.C. Available: http://www.justice.gov/atr/public/speeches/289708.pdf.

Shaohua, Shi. 2007. Introduction of the CESI IPR Policy Template. Standardization Development Research Center, China Electronics Standardization Institute. Available: http://docbox.etsi.org/workshop/2007/2007_IPR_Symposium/012.1%20EN%20-%20CESI_SHI%20SHAOHUA.pdf.

Sharma, C. 2012. Global Mobile State of the Union 2012. Available: http://www.chetansharma.com/research.htm.

Spence, A. M. 1973. Job Market Signaling. Quarterly Journal of Economics. 87: 355-374.

Standards Administration of China. 2013. Regulatory Measures on National Standards Involving Patents (Interim). Available: http://sunsteinlaw.com/wp/wp-content/uploads/2013/01/2013_01_IP_Update_PRC.pdf.

Suttmeier, R. and X. Yao. 2011. China's IP Transition: Rethinking Intellectual Property Rights in a Rising China. National Bureau of Asian Research. Available: http://www.nbr.org/publications/issue.aspx?id=232.

Swanson, D. and W. Baumol. 2005. Reasonable and Nondiscriminatory (FRAND) Royalties, Standards Selection, and Control of Market Power. Antitrust Law Journal 73(1):1-58.

Tapia, C 2010. Industrial Property Rights, Technical Standards and Licensing Practices (FRAND) in the Telecommunications Industry. Germany: Heymanns Verlag GmbH.

Torrance, A. and L. Kahl. 2012. Synthetic Biology Standards and Intellectual Property. Available: http://sites.nationalacademies.org/PGA/step/PGA_058712.

U.S. Department of Justice. 2006. Business Review Letter to VMEbus International Trade Association. Available: http://www.justice.gov/atr/public/busreview/219380.htm.

U.S. Department of Justice. 2007. Business Review Letter to Institute of Electrical and Electronics Engineers. Available: http://www.justice.gov/atr/public/busreview/222978.pdf.

U.S. Federal Trade Commission. 2011. The Evolving IP Marketplace: Aligning Patent Notice and Remedies with Competition. Available: http://www.ftc.gov/os/2011/03/110307patentreport.pdf.

U.S. Federal Trade Commission Press. 2011. Press Release: FTC Order Restores Competition in U.S. Market for Equipment Used to Recharge Vehicle Air Conditioning Systems: Under Settlement, Bosch Agrees to Sell RTI Brand to Mahle Clevite, Inc. and to Make Certain Patents Available to Competitors. Available: http://www.ftc.gov/opa/2012/11/bosch.shtm.

U.S. Department of Justice and Federal Trade Commission. 2007. Antitrust Enforcement and Intellectual Property Rights: Promoting Innovation and Competition. Available: http://www.justice.gov/atr/public/hearings/ip/222655.pdf.

U.S. Department of Justice and United States Patent and Trademark Office. 2013. Policy Statement on Remedies for Standard-Essential Patents Subject to Voluntary F/RAND Commitments. Available: http://www.uspto.gov/about/offices/ogc/Final_DOJ-PTO_Policy_Statement_on_FRAND_SEPs_1-8-13.pdf.

USITO. 2007. Chinese ICT Standards Landscape. United States Information Technology Office, Beijing. Issue Paper No. 621. Available: http://read.pudn.com/downloads141/doc/607628/chinese-ICT-standards.pdf.

Ward, J. 2012. Revenue for Mobile Communications Equipment to Climb in the Double Digits by Year End: Smartphones, Tablets, and 4G LTE Help Move Forward in Spite of Soft Global Economy. IHS iSuppli Market Research. Available: http://www.isuppli.com/Mobile-and-Wireless-Communications/MarketWatch/pages/Revenue-for-Mobile-Communications-Equipment-to-Climb-in-the-Double-Digits-by-Year-End.aspx.

Wayland, J. 2012. Oversight of the Impact of Competition of Exclusion Orders to Enforce Standards-Essential Patents. Department of Justice, Washington, D.C. Available: http://www.justice.gov/atr/public/testimony/284982.pdf.

Willingmyre, G. 2009. China's Proposed Regulations for Patent-Involving National Standards. Available: http://www.ip-watch.org/2009/12/21/take-two-china%E2%80%99s-proposed-regulations-for-patent-involving-national-standards/.

Willingmyre, G. 2010. China's Latest Draft Proposal Rules for Patents in Standards a Step Forward? Available: http://www.ip-watch.org/2010/04/01/china%E2%80%99s-latest-draft-disposal-rules-for-patents-in-standards-a-step-forward/.

World Intellectual Property Organization. (as amended by the Law of July 31, 2009). Germany's Patent Law Section 15 (3). Available: http://www.wipo.int/wipolex/en/text.jsp?file_id=238776.

Yeh, B. 2012. Availability of Injunctive Relief for Standard-Essential Patent Holders. Congressional Research Service, Washington, D.C.

JUDICIAL CASES

A.C. Aukerman Company v. R.L. Chaides Construction Co., 960 F.3d 1020, 1032 (Fed. Cir. 1992).
Apple Inc. v. Motorola, Inc., 869 F. Supp. 2d 901 (N.D. Ill. June 22, 2012).
Apple Inc. v. Motorola, Inc., Nos. 2012-1548, 2012-1549 (Fed. Cir.).
Association for Molecular Pathology et al. v. Myriad Genetics, Inc. et al. (Supreme Court no. 12-398, 2013).
In re Dell Corporation, Dell VESA case, No. C-3658. 121 F.T.C. 616 (May 20, 1996).
eBay v. MercExchange, L.L.C., 547 U.S. 388 (2006).
Fujitsu Ltd. v. Netgear Inc., 2010 U.S. App. LEXIS 26722 (Fed. Cir., Nov. 1, 2010).

References

Georgia-Pacific Corp. v. United States Plywood Corp, 318 F. Supp. 1116 (S.D.N.Y. 1970).

Huawei v. ZTE, 2013, Regional Court of Düsseldorf, Federal Republic of Germany, Case No. 4b O 104/12.

In re Proxim Corporation, et al., Debtors, In the United States Bankruptcy Court for the District of Delaware, Case No. 05-11639 (PJW) (Jointly Administered) (Chapter 11). Available: http://www.ftc.gov/os/caselist/0511639/0511639.shtm.

In the Matter of Google Inc., FTC File No. 121-0120. (Jan. 3, 2013). Separate Statement of Commissioner J. Thomas Rosch Regarding Google's Standard-Essential Patents Enforcement Practices. No. C-Available: http://www.ftc.gov/os/caselist/1210120/1301 03googlemotorolaroschstmt.pdf.

Illinois Tool Works Inc., et al., Petitioners v. Independent Ink, Inc., 547 U.S. 126 (2006).

IPCom v. Nokia and HTC (2012) EWCA Civ 56.

Koninklijke Philips Electronics N.V. v. SK Kassetten GmbH & Co. District Court The Hague, The Netherlands, 17 March 2010, Joint Cases No. 316533/HA ZA 08-2522 and 316535/HA ZA 08-2524.

KSR International. Co. v. Teleflex, Inc. 550 U.S. 398 (2007).

Motorola v. Microsoft, Regional Court of Mannheim, Federal Republic of Germany, Case No. 2 O 240/11. (2012).

Microsoft Corp. v. Motorola, Inc., No. C10-1823JLR, 2012 U.S. Dist. LEXIS 170587 (W.D. Wash. Nov. 30, 2012).

Microsoft Corp. v. Motorola, Inc., 871 F. Supp. 2d 1089 (W.D. Wash. 2012), *aff'd*, 696 F.3d 872 (9th Cir. 2012).

In the Matter of Motorola Mobility LLC, FTC File No. 121-0120, No. C- (Jan. 3, 2013). http://www.ftc.gov/os/caselist/1210120/130103googlemotorolado.pdf.

Motorola v. Apple, 2012, Higher Regional Court of Karlsruhe, Federal Republic of Germany, Case No. 6 U 136/11.

In re Negotiated Data Solutions LLC, FTC File No. 051-0094. Available: http://www.ftc. gov/os/caselist/0510094/080122do.pdf.

Pratt v. Wilcox Mfg. Co., 64 F. 589 (N.D. Ill. 1893).

Qualcomm, Inc. v. Broadcomm, Inc., No. 2007-1545 (Fed. Cir. Dec. 1, 2008).

Radio Systems Corp. v. Lalor, No 2012-123 (Fed. Cir. 2013).

Rambus Inc. v. Infineon Tech. AG., 318 F. 3d. 1081, 1097 (Fed. Cir. 2003).

Rembrandt Data Techs., LP v. AOL, LLC et al., Case No. 10-1002 (Fed. Cir. 2011).

In the Matter of Robert Bosch GmbH, No. C-4377 (Nov. 26, 2012).

Samsung v. Apple, District Court of The Hague, 20 June 2012, case numbers/docket numbers 400367/HA ZA 11-2212, 400376/HA ZA 11-2213 and 400385/HA ZA 11-22 15.

Spindelfabrik Suessen-Schurr Stahlecker & Grill v. Schubert & Salzer Maschinenfabrik AG, 829 F.2d 1075 (Fed. Cir. 1987).

SRI International Inc v. Internet Security Systems Inc., 511 F.3d 1186 (Fed. Cir. 2008).

TransCore LP v. Electronic Transaction Consultants Corporation, 563 F3d 1271(5th Cir 2009).

Wang Laboratories Inc. v. Mitsubishi Electronics America, Inc., 103 F.3d 1571 (Fed. Cir. 1997).

Case No COMP/M.6381 Google/Motorola Mobility, Regulation (EC) No 139/2004 Merger Procedure (2012) Available: http://ec.europa.eu/competition/mergers/cases/de cisions/m6381_20120213_20310_2277480_EN.pdf.

Third Party United States Federal Trade Commission's Statement on the Public Interest. *In the Matter of Certain Gaming and Entertainment Consoles, Related Software, and*

Components Thereof, Recommended Determination on Remedy and Bonding (May 18, 2012). Inv. No. 337-TA-752. Available: http://www.ftc.gov/os/2012/06/1206 ftcgamingconsole.pdf.

Third Party United States Federal Trade Commission's Statement on the Public Interest. *In the Matter of Certain Wireless Communication Devices, Portable Music and Data Processing Devices, Computers and Components Thereof*, Notice Regarding Initial Determination on Violation of Section 337. (April 24, 2012). Inv. No. 337-TA-745. Available: http://www.ftc.gov/os/2012/06/1206ftcwirelesscom.pdf.

Appendix A

Acronyms

3GPP	Third Generation Partnership Project
ABNT	Brazilian Association of Technical Norms
ABS	Academy of Broadcasting Science (China)
ADR	alternative dispute resolution
AML	anti-monopoly law
ANSI	American National Standards Institute
AQSIQ	General Administration of Quality Supervision, Inspection, and Quarantine (China)
AVS	Audio Video Coding Working Group of China
BIS	Bureau of Indian Standards
CADE	Competition Administrative Court (Brazil)
CAS	China Association for Standardization
CATR	China Academy of Telecommunications Research
CCSA	China Communications Standards Association
CESA	China Electronics Standards Association
CESI	China Electronic Standardization Institute
CIPO	Chinese State Intellectual Property Office
CJEU	Court of Justice of the European Union
CNIS	China National Institute of Standardization
CONMETRO	National Council of Metrology, Standardization, and Industrial Quality (Brazil)
COSTIND	Commission of Science, Technology and Industry for National Defense (China)
DOJ	Department of Justice
DOSTI	Development Organization of Standards for Telecommunications in India
EPO	European Patent Office
ETSI	European Telecommunications Standards Institute
FRAND	fair, reasonable and non-discriminatory
FTC	Federal Trade Commission
GISFI	Global ICT Standardization Forum for India

HDMI	High Definition Multimedia Interface
IEC	International Electrotechnical Commission
IEEE	Institute of Electrical and Electronics Engineers
IEEE-SA	Institute of Electrical and Electronics Engineers Standards Association
IETF	Internet Engineering Task Force
ICT	information and communications technology
INMETRO	National Institute of Metrology, Standardization, Quality and Technology (Brazil)
INPI	Brazil Patent and Trademark Office, National Industrial Property Institute (Brazil)
ISO	International Organization for Standardization
ITC	International Trade Commission
ITU	International Telecommunications Union
ITU-T	International Telecommunications Standardizations Sector
ISO	International Organization for Standardization
IPR	intellectual property rights
LITD	Electronics and Information Technology Division Council (India)
MIIT	Ministry of Industry and Information Technology (China)
MII	Ministry of Information Industry (China)
MOST	Ministry of Science and Technology (China)
MOU	memorandum of understanding
NCAC	National Copyright Administration of China
NDRC	National Development and Reform Commission (China)
NFC	Nearfield Communications Forum
NIST	National Institute of Standards and Technology
NPL	non-patent literature
OASIS	Organization for the Advancement of Structured Information Standards
RAND	reasonable and non-discriminatory
RF	royalty free
SAC	Standards Administration of China
SAIC	State Administration for Industry and Commerce (China)
SANBFT	State Administration for Broadcasting, Film, and Television (China)
SARFT	State Administration for Radio, Film, and Television (China)
SASTIND	State Administration for Science, Technology and Industry (China)
SCITO	State Council Informatization Office (China)
SDO	standards development organization
SEP	standard-essential patent
SGIP	smart grid interoperability panel
SIG	special interest group

Appendix A

SINMETRO	National System of Metrology, Standardization, and Industrial Quality (Brazil)
SIPO	State Intellectual Property Office (China)
SSO	standard-setting organization
STIC	Scientific and Technical Information Center (U.S.)
TBT	Technical Barriers to Trade
TEC	Telecommunications Engineering Center (India)
TRIPs	Trade-Related Intellectual Property Rights
TSDSI	Telecommunications Standards Development Society (India)
USITO	United States Information Technology Office
USPTO	United States Patent and Trademark Office
VITA	VMEBus International Trade Association
WIPO	World Intellectual Property Organization
WTO	World Trade Organization
W3C	World Wide Web Consortium

Appendix B

Symposium Agenda

Symposium on
Management of Intellectual Property in Standard-Setting Processes

October 3-4, 2012
NAS Building, Lecture Room
2101 Constitution Ave., NW, Washington, DC

Wednesday, October 3
Part One: Institutional and National Diversity

8:30   **Welcome and Introduction:** Keith Maskus, University of Colorado

8:45   **Keynote:** Stuart Graham, U.S. Patent and Trademark Office

9:15   **Session 1: Policies, Practices, and Experience of Leading Standards Organizations**
Chair: Tim Simcoe, Boston University
Presentation: Rudi Bekkers, Eindhoven University of Technology, Netherlands and Andy Updegrove, Gesmer Updegrove, LLP
Discussants:
Fiona Scott-Morton, U.S. Department of Justice
Dirk Weiler, European Telecommunications Standards Institute (ETSI)
John Kelly, JEDEC Solid State Technology Association

10:45   Break

11:00   **Session 2: Standards Processes and IP Treatment in Emerging Economies**
Chair: Richard Suttmeier, University of Oregon, ret.

Presentations:
China – Danny Breznitz, Georgia Institute of Technology
India – Thammaiah Ramakrishna, National Law University, Bangalore, India
Brazil – Denis Barbosa, Catholic University of Rio de Janeiro, Brazil
Discussants:
Julia Doherty, Office of the U.S. Trade Representative
Mark Cohen, Fordham University Law School
Kent Baker, Consultant

12:30 Lunch

1:30 Session 3: E-government Procurement Policies in the US, EU, and Japan
Chair: Amy Marasco, Microsoft
Presentation: Laura DeNardis, American University
Discussants
Naomi Voegtli, SAP
Mary Saunders, National Institute of Standards and Technology

3:00 Break

Part Two: II. Structural and Policy Issues in IP Management in the Standards Context

3:15 Session 4: SDO-Patent Office Cooperation and Information Sharing
Chair: Rudi Bekkers, Eindhoven University of Technology, Netherlands
Panelists: Michel Goudelis, European Patent Office
Dirk Weiler, European Telecommunications Standards Institute
George Willingmyre, GTW Associates

4:45 Session 5: Standards-Essential Patents and USITC Litigation

Chair: David Goodman, Polytechnic Institute of New York University
Presentations:
Colleen Chien, University of Santa Clara Law School
Richard Gilbert, University of California at Berkeley
Discussant: Suzanne Munck, Federal Trade Commission

6:00 Adjourn

Appendix B

Thursday, October 4, 2012

8:30 Introduction: Keith Maskus, University of Colorado
Keynote: Howard Shelanski, Federal Trade Commission

9:00 **Session 6: Standards Development in Emerging Technologies**
Chair: Richard Gilbert, University of California at Berkeley
Presentations:
Bioinformatics – Jorge Contreras, American University Law School
Nanotechnology – Ajit Jillavenkatesa, National Institute of Standards and Technology
Synthetic biology – Andrew Torrance, University of Kansas
Green building materials – Jorge Contreras, American University Law School
Discussant:
Arti Rai, Duke University Law School

10:30 Break

10:45 **Session 7: Transfer of Patents and Obligations**
Chair: Sandy Block, IBM
Presentation: Jay Kesan, University of Illinois Law School
Discussants:
Gil Ohana, Cisco
Scott Peterson, Google
Claudia Tapia, Research in Motion

12:15 **Open Forum***
Chair: Ollie Smoot, Past President, International Organization for Standardization
Comments:
Monica Barone, American Intellectual Property Law Association (AIPLA)
Dan Bart, Valley View Corporation
Carter Eltzroth, Helikon.net
Keith Mallinson, WiseHarbor
Tim Molino, Business Software Alliance (BSA)
Ian McClure, Intellectual Property Exchange International (IPXI)
Brian Pomper, Akin Gump Strauss Hauer & Feld LLP
*This session is intended to give stakeholders an opportunity to comment further on issues on the agenda or raise new issues regarding IP in standards for the committee's consideration.

1:15 **Closing Remarks:** Keith Maskus

1:30 **Adjourn**

Appendix C

Biographies of Committee and Staff

Keith E. Maskus, *Chair,* is Professor of Economics at the University of Colorado, Boulder. He has been a Lead Economist in the Development Research Group at the World Bank. Dr. Maskus is also a Research Fellow at the Peterson Institute for International Economics, a Fellow at the Kiel Institute for World Economics, and an Adjunct Professor at the University of Adelaide. He has been a visiting professor at the University of Bocconi, and a visiting scholar at the Center for Economic Studies-Ifo Institute at the University of Munich and the China Center for Economic Research at Peking University. He also serves as a consultant for the World Bank, the World Health Organization, and the World Intellectual Property Organization. Dr. Maskus received his Ph.D. in Economics from the University of Michigan in 1981 and has written extensively about various aspects of international trade. His current research focuses on the international economic aspects of protecting intellectual property rights. He is the author of *Intellectual Property Rights in the Global Economy*, published by the Institute for International Economics, and co-editor of *International Public Goods and the Transfer of Technology under a Globalized Intellectual Property Regime*, published by Cambridge University Press. He recently wrote a piece analyzing the need for reforms in U.S. patent policy, published by the Council on Foreign Relations.

Rudi Bekkers is a tenured faculty member at the Eindhoven University of Technology, Netherlands and specializes in the relationship between standardization and intellectual property rights. Over the last 15 years, he has published a number of papers on this topic in established journals. In addition, he performed more than a dozen commissioned studies and projects on standards for the European Commission, the Organization for Economic Cooperation and Development (OECD), various national ministries, standards bodies such as the European Standards Telecommunications Institute (ETSI), and for companies and other stakeholders. His recent projects include a fact-finding study on intellectual property rights in standards, commissioned by the European Commission. Executed in 2010-2011, this study included a quantitative study of disclosed intellectual property rights for standards, and considered how design aspects of IPR policies affect an efficient and well performing market. Currently, Dr. Bekkers is collaborating with committee member Timothy Simcoe, University of Boston, to create a comprehensive, a public database of IPR disclosures at standards bodies, a project that was first announced at the a National Bureau of Economic Research (NBER) preconference on Standards, Patents and Innovation in May 2011.

Marc Sandy Block is IP Counsel at IBM working in standards, bankruptcy, and IP policy. For a period, Sandy managed the corporation's Latin American patent portfolio and also managed several intellectual property law departments. For several years, Mr. Block was President of the International Intellectual Property Society (iipsny.org) and was a board member for over ten years. He was a contributor to the American Bar Association Manual on Standards and Development and has been a member of the ANSI IPR Policy Committee, AIPLA Standards and Open Source Committee, and IPO Standards Committee. He is also a guest lecturer at Cardozo Law School and has spoken and published articles in the fields of IP, standards, and bankruptcy. He recently participated as a panelist at the FTC Workshop on Patent Issues in Standards. Prior to IBM, he was a patent attorney at Hall, Myers and Rose in Potomac, Maryland, and Washington, DC. Also prior to IBM, a programmable implantable medication infusion system patent that he prepared was named a 1984 Intellectual Property Owners (IPO) Invention of the Year. In a previous incarnation, he was an officer in a Special Forces Signal Company (11th SF Group USAR) in Ft. Meade, Maryland. Mr. Block earned a B.S. in Electrical Engineering from Lehigh University and a J.D. from the George Washington University Law School.

Jorge L. Contreras is an Associate Professor of Law at American University's Washington College of Law where he teaches intellectual property and property law. Previously, he served as a Senior Lecturer and Acting Director of the Intellectual Property Program at Washington University in St. Louis, School of Law. His research focuses on the effects of intellectual property structures and governmental regulation on the dissemination of scientific and technological innovation. Professor Contreras also serves as co-chair of the American Bar Association's Section of Science and Technology Law Committee on Technical Standardization and co-chair of the National Conference of Lawyers and Scientists. Professor Contreras is the editor of the American Bar Association's Technical Standards Patent Policy Manual (ABA Publishing: Chicago, 2007) and has published numerous articles and book chapters relating to legal issues surrounding intellectual property, scientific research, and standards development. He has served as the principal legal counsel to the Internet Engineering Task Force (IETF), the leading developer of Internet architecture, transport and security standards, since 1998 and is a Fellow of the American Bar Foundation. Prior to entering academia Professor Contreras was a partner at the international law firm Wilmer Cutler Pickering Hale and Dorr LLP. Professor Contreras holds

Bachelor of Arts and Bachelor of Science in Electrical Engineering degrees from Rice University and a Juris Doctor degree from Harvard Law School.

Richard Gilbert is Emeritus Professor of Economics and Professor of the Graduate School at the University of California at Berkeley. He was Chair of the Department of Economics at Berkeley from 2002 to 2005 and is currently Chair of the Berkeley Competition Policy Center. From 1993 to 1995, he was Deputy Assistant Attorney General in the Antitrust Division of the U.S. Department of Justice where he led the effort that developed joint Department of Justice and Federal Trade Commission Antitrust Guidelines for the Licensing of Intellectual Property. Before serving in the Department of Justice, Dr. Gilbert was the Director of the University of California Energy Institute and Associate Editor of the Journal of Industrial Economics, the Journal of Economic Theory, and the Review of Industrial Organization. He is a former President of the Industrial Organization Society. His research specialties are in the areas of competition policy, intellectual property, and research and development. He has lectured widely and testified in proceedings before state and federal courts, regulatory commissions, the California Legislature, and the U.S. Congress. Dr. Gilbert holds a Ph.D. in Engineering-Economic Systems from Stanford University and Bachelor of Science and Master of Science degrees in Electrical Engineering from Cornell University.

David Goodman is a member of the National Academy of Engineering and a foreign member of the Royal Academy of Engineering, a Fellow of the Institute of Electrical and Electronic Engineers, and a Fellow of the Institution of Engineering and Technology. He retired from his position as Professor of Electrical and Computer Engineering and Director of the Wireless Internet Center for Advanced Technology (WICAT) at Polytechnic Institute of New York University in June 2008. David's research has made fundamental contributions to digital signal processing, speech coding, and wireless information networks. He received the ACM/SIGMOBILE Award for "Outstanding Contributions to Research on Mobility of Systems Users, Data, and Computing" in 1997 and the Avant Garde award from the Vehicular Technology Society of the IEEE in 2003. In 1997, David served as Chairman of the National Research Council Committee studying The Evolution of Untethered Communications. He has also worked as Program Director in the Computer and Network Systems Division of the National Science Foundation (2006-2007), Head of the Electrical and Computer Engineering Department at Polytechnic Institute (1999-2001), and Research Associate at the Program on Information Resources Policy at Harvard University (1995). He is author and co-editor of several other books on wireless communications. David received a Bachelor's degree at Rensselaer Polytechnic Institute (1960), a Master's at New York University (1962), and a Ph.D. at Imperial College, University of London (1967), all in Electrical Engineering.

Amy Marasco is the General Manager for Standards Strategy and Policy at Microsoft. She leads a team that addresses strategic policy and engagement issues on a corporate-wide, global basis. Ms. Marasco regularly engages in policy discussions involving standards, intellectual property rights and competition law issues at numerous standards bodies and in many other forums. She is a rapporteur at the ITU-T IPR Ad Hoc Group, a co-Chairman of the Standards Policy Committee at the Intellectual Property Owners Association, and Chairman of Standards and Intellectual Property Rights Policy Committee at the Telecommunications Industry Association. She has testified or given presentations on standards-related policy issues upon request by the U.S Federal Trade Commission and U.S. Department of Justice (Antitrust Division), the European Commission, the Japanese Ministry of Economy, Trade and Industry (METI), China Electronic Standards Institute (CESI) and the China National Institute of Standardization (CNIS) in the People's Republic of China. Ms. Marasco joined Microsoft after serving as the Vice President and General Counsel of the American National Standards Institute (ANSI) from 1994-2004. Prior to joining ANSI, she was an attorney with the law firm of Cadwalader, Wickersham & Taft in its New York office.

Timothy Simcoe is Assistant Professor of Strategy and Innovation at Boston University School of Management. He is also a Faculty Research Fellow at the National Bureau of Economic Research and an Associate Editor of Management Science. Dr. Simcoe's research specialities are the economics of innovation, science and technology policy, intellectual property and corporate strategy. He has published numerous articles and book chapters on intellectual property and standards development. Dr. Simcoe received an A.B. in Applied Math from Harvard University, an M.A. in Economics from the University of California at Berkeley, and a Ph.D. in Business Administration from the University of California at Berkeley.

Oliver R. Smoot is a consultant on standards and intellectual property issues. He served as Chairman of the American National Standards Institute (ANSI) Board of Directors in 2001 and 2002, and past Chairman from 2003-2005. Before being elected as Chairman of the ANSI Board, Mr. Smoot served in numerous ANSI leadership posts, including Chair of ANSI's Finance Committee, Organizational Member Council, and Patent Group. In 2003-2004, Mr. Smoot was elected for a two-year term as President of the International Organization for Standardization (ISO), a worldwide federation of national standards bodies with representatives from over 150 countries. From 2000-2005, Mr. Smoot served as Vice-President for External Voluntary Standards Relations of the Information Technology Industry Council (ITI). Prior to that appointment Mr. Smoot was ITI's executive vice-president for 23 years. An active member of the American Bar Association for many years, Mr. Smoot served as chairman of the Section on Science and Technology Law and most recently as chairman of its Technical Standardization Law Committee. He has also served in numerous positions with the Computer Law Association (now the International Technology Law Association), culminating as President. Mr. Smoot currently serves on the Executive Committee of the U.S. Policy Committee of the Association for Computing Machinery. He received a B.S. from MIT and a J.D. from Georgetown University.

Richard P. Suttmeier is a Professor of Political Science, Emeritus, at the University of Oregon. He has written widely on science and technology development issues in China. His most recent publications include: "China's IP Transition: Rethinking Intellectual Property Rights in a Rising China" (with Yao Xiangkui) (National Bureau of Asian Research. July, 2011) and "Standards, Stakeholders, and Innovation: China's Evolving Role in the Global Knowledge Economy" (with Scott Kennedy and Jun Su) (National Bureau of Asian Research. September, 2008). Dr. Suttmeier has served as Senior Analyst, Congressional Office of Technology Assessment, and Director of the Beijing Office, National Academy of Sciences/Committee on Scholarly Communication with the People's Republic of China, and as a consultant to the World Bank, the UNDP, and the U.S. government. He recently completed several months service as Senior Visiting Scholar at the Institute of Policy and Management of the Chinese Academy of Sciences.

Andrew Updegrove is a founding partner of Gesmer Updegrove LLP, a Boston-based technology law firm. He has a broad range of experience in representing both mature and emerging high technology companies of all types in all aspects of their legal affairs. Since 1988, he has also represented and helped structure more than 110 worldwide standard-setting, open source, research and development, promotional and advocacy consortia, including some of the largest standard setting organizations in the world. In 2005 he was elected to the Boards of Directors of the American National Standards Institute (ANSI) and to the Free Standards Group (FSG). In 2007 he was elected to the Board of Directors of the Linux Foundation. He is a current member of the Board of Advisors of Open Source for America, and a Charter Fellow of the OpenForum Academy. Mr. Updegrove has also provided testimony to the Department of Justice and Federal Trade Commission on standard-setting and intellectual property rights, and written and filed pro bono "friend of the court" briefs in major standards-related litigation before the Federal Circuit Court, the Supreme Court, and the Federal Trade Commission. In 2002, he conceived and launched ConsortiumInfo.org, an extensive website intended to provide the most comprehensive and detailed source of news and information on standard-setting, open source software project development, and forming and maintaining consortia. Mr. Updegrove is a graduate of Yale University and the Cornell University Law School.

Staff

Stephen A. Merrill, project director, has been Executive Director of the National Academies' Board on Science, Technology, and Economic Policy (STEP) since its formation in 1992. With the sponsorship of numerous federal government agencies, foundations, multinational corporations, and international institutions, the STEP program has become an important discussion forum and authoritative voice on innovation, competitiveness, intellectual property, human resources, statistical, and research and development policies. At the same time Dr. Merrill has directed many STEP projects and publications, including *A Patent System for the 21st Century* (2004), *Innovation Inducement Prizes* (2007), and *Innovation in Global Industries* (2008). For his work on patent reform he was named one of the 50 most influential people worldwide in the intellectual property field by *Managing Intellectual Property* magazine and earned the Academies' 2005 Distinguished Service Award. He has been a member of the World Economic Forum Global Council on the Intellectual Property System. Previously, Dr. Merrill was a Fellow in International Business at the Center for Strategic and International Studies (CSIS), where he specialized in technology trade issues. He served on various congressional staffs including the U.S. Senate Commerce, Science, and Transportation Committee, where he organized the first congressional hearings on international competition in the semiconductor and biotechnology industries. Dr. Merrill holds degrees in political science from Columbia (B.A.), Oxford (MPhil), and Yale (M.A. and Ph.D.) Universities. He attended the Kennedy School of Government's Senior Executives Program and was an adjunct professor of international affairs at Georgetown University from 1989 to 1996.

Aqila Coulthurst has been Program Coordinator for STEP since the fall of 2011. Prior to joining STEP, she spent over two years in the production and marketing divisions of the National Academies Press (NAP). Ms. Coulthurst was involved in several initiatives at NAP including: direct marketing and online outreach; facilitating the sale of intellectual property rights to publishers abroad, and general operational support. Over the years, Ms. Coulthurst has worked in various capacities at Smithsonian Enterprises, the National Community Action Foundation, Kingsley Associates and the Center for Science, Technology and Economic Development at SRI International. She has extensive experience conducting impact assessments and program evaluations.

In addition to her interest in U.S. competitiveness and innovation policies, Ms. Coulthurst is interested in how these policies impact development abroad. She spent several years studying U.S. foreign policy and sustainable development at renowned institutions in Washington and while studying abroad in Central America. She has a B.A. in economics and in Spanish, and a certificate in markets and management from Duke University. She also has a Master of Science in Foreign Service from Georgetown University.